POLYMER ENGINEERING

H. LEVERNE WILLIAMS

Department of Chemical Engineering and Applied Chemistry,
Faculty of Applied Science and Engineering, University of Toronto,
Toronto, Ontario, Canada

ELSEVIER SCIENTIFIC PUBLISHING COMPANY
AMSTERDAM — OXFORD — NEW YORK   1975

ELSEVIER SCIENTIFIC PUBLISHING COMPANY
335 JAN VAN GALENSTRAAT
P.O. BOX 211, AMSTERDAM, THE NETHERLANDS

AMERICAN ELSEVIER PUBLISHING COMPANY, INC.
52 VANDERBILT AVENUE
NEW YORK, NEW YORK 10017

ISBN 0-444-41381-2

Printed in The Netherlands

)

620. 192

WIL

to Molly

Contents

Preface

Chemical Engineers have been accused of being jacks-of-all-trades and masters-of-none, which one can believe in the sense that Chemical Engineers find their ways into the most diverse of industries and positions. However, they are quite capable of mastering the relevant engineering and technological aspects of those industries. The problem lies in the fact that during the few years at University it is impossible to become expert in every conceivable field of interest.

Thus there is more tendency for the undergraduate Chemical Engineer to be trained as a generalist and to specialize, perhaps in the senior year but more likely, during graduate studies. Since such a large number of Chemical Engineers are employed by the polymer-based industries it is only fitting that they should have at least a working knowledge of polymer engineering in preparation for their duties.

The subject falls logically into three aspects. Firstly there is monomer synthesis and conversion to polymers. This industry is so closely related to the chemical and petroleum industry that most students will have received an adequate background in this subject, if not in detail, at least as far as principles are concerned so that they can read with profit the excellent monographs available. Likewise at the other end of the scale there is the technology of handling and fabrication which is so diverse and so closely related to the individual Company that detailed training in this subject is best done on the job.

However, in between there are a number of concepts which enable one to translate knowledge of how the polymer is prepared into at least some idea of how it will perform under technological conditions. The subject is enormous but the average student can be allowed only one course for one semester, say 25 lectures in all, for this. The following monograph was compiled to fit into this scheme and has been used for several

years both for senior students and graduate students, usually
having no prior polymer chemistry or polymer technology and
often without training desirable in organic chemistry or
physics.

Several excellent texts can be used as supplementary reading.
However an exhaustive point-by-point reference to the many
eminent scientists and engineers who pioneered the studies has
been omitted since students seldom make use of these. Rather,
readily available and recent texts are recommended for
additional reading and specific papers are mentioned only when
the contents are not likely to be found easily in texts. Likewise elaborate and scientifically sound diagrams have been
left out in favor of general diagrams illustrating the ideas.
The fine points, arguments, mathematical derivations, and side
issues again are left for outside reading.

The purpose of the text is to give to the young Chemical
Engineer entering industry some idea of what a polymer is, what
it will do, and how it fails so that whether his position is
one in management, sales, production, or research he has a
basis from which to work, a guide map for the polymer globe
and a starting point for more detailed travels guided by the
more extensive, and expensive, monographs now available.

Finally as a start for almost any survey the relevant sections of the "Encyclopedia of Polymer Science and Technology",
Wiley-Interscience, can be recommended.

H. Leverne Williams

Toronto, 1975.

1

Introduction: The Nature of High Polymers

This monograph is intended to cover the triangle structure-properties-uses of high polymers. The subject is vast but one with which more and more Chemical Engineers need to be well acquainted. To maintain the length of the text within reason, it must be assumed that the reader has prior knowledge of the synthesis of polymers, the organic chemistry, and the characterization of high polymers, the physical chemistry. However, this may not be true in every case and this introductory chapter is intended to summarize polymer science either as a reference or as a refresher for the subsequent chapters which treat polymers as materials with minimal reference to specific structural details and chemistry.

Polymers form such a large and diverse group of materials that a precise definition is impossible. Natural polymers include proteins, starchs, celluloses, lignins, silk, cotton, wool, deoxyribonucleic acids (DNA) and ribonucleic acids (RNA). Synthetic polymers include polyolefins, polyamides, epoxy resins, polyesters, polyurethanes and polyethers known familiarly as plastics, elastomers, fibers and resins. They are used in fabricating plastic and elastic articles, adhesives, coatings, sealants, fibers, films, fabrics, panels, and the host of other objects in everyday use. Usually one can define a polymer as a giant molecule, high polymer, or macromolecule made of many (poly) units (mers) which repeat along the molecule. The single units (monomers) may be all the same as in the case of the homopolymers, cellulose and polyethylene, or there may be at least two as in the case of the copolymers, proteins and epoxy resins.

Most natural and synthetic polymers arise logically from organic chemistry, the chemistry of the compounds of carbon, and depend upon the ability of carbon atoms to bond covalently

2

with each other as well as with other atoms. This leads to
long chains the simplest of which are composed of carbon and
hydrogen having the general formula;

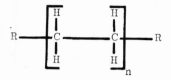

in which (n) is a large number and can vary indeed from one,
i.e. ethane, to many thousands as in polyethylene. Within a
single molecule the carbon and hydrogen atoms are joined covalently and the individual molecules interact by intermolecular forces such as van der Waals, hydrogen-bonding and other
secondary forces which will be mentioned in context.

One other very important fact must be recalled about the
carbon atom in its compounds. Normally it is present as
tetravalent carbon with four equivalent SP^3 bonds (solid
lines) pointing towards the four corners of a tetrahedron
(dotted lines) hence in practice the polyethylene

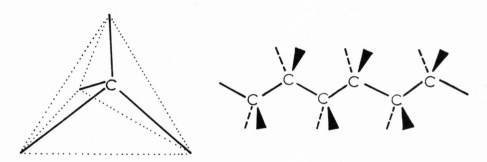

chain would be a zig-zag or spiral. Also carbon may be present
in so-called unsaturated double or triple bond forms usually
designated

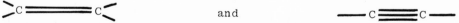

        and

but one must not forget that the carbon is still tetravalent,
but bonds are three SP^2 and one P and two SP and two P respectively. The P bonds represent "unsaturation" in the sense
that they are easily converted to normal covalent bonds by the

addition of hydrogen, leaving one SP^3 bond between the carbon
atoms. Details of the bonding of the carbon atom will be found
in numerous texts on organic chemistry. For the purposes of
this discussion it is only necessary to realize that normally
carbon yields a tetrahedral structure which becomes planar
about a

C⚌⚌⚌C   and linear about a   C⚌⚌⚌C

bond, and further that these two unsaturated structures tend
to revert to the saturated structures upon reaction with hydrogen or its equivalent; chlorine, bromine, and other reactive
elements and compounds. We cannot review at this time all the
possible structures when other atoms such as oxygen, nitrogen,
sulfur, silicon, phosphorus, or boron are included in the
polymer molecule. Suffice to point out at this stage that the
strength, direction, and length of the bonds between atoms in
a single molecule govern such properties as thermal stability,
rigidity, electrical properties, and optical properties whereas
the weaker intermolecular forces are important in solubility,
rheology (flow), viscoelastic properties, crystallization, and
orientation.

Classification of Polymers

There is no precise method of classifying polymers but
rather there are a number of groups into which they may be
placed for convenience. The first is based on the chemistry
of the synthesis [1,2,3,4]. If, for example, ethylene gas is
mixed with an initiator which yields free radicals, i.e. a
reactive species with one unshared electron, initiation of
polymerization will take place:

Initiator   R:R $\longrightarrow$ 2R·   Free Radicals

R· + [ethylene structure] $\longrightarrow$ R—CH$_2$—CH$_2$·   Initiation

This is followed by propagation during which the chain rapidly
grows in length, hence the term chain reaction, or free

4

radical chain reaction.

$$R-\overset{\overset{\displaystyle H}{|}}{\underset{\underset{\displaystyle H}{|}}{C}}-\overset{\overset{\displaystyle H}{|}}{\underset{\underset{\displaystyle H}{|}}{C}}\cdot \;+\; n\;\; \overset{\overset{\displaystyle H}{|}}{\underset{\underset{\displaystyle H}{|}}{C}}=\overset{\overset{\displaystyle H}{|}}{\underset{\underset{\displaystyle H}{|}}{C}} \longrightarrow R-\left[\overset{\overset{\displaystyle H}{|}}{\underset{\underset{\displaystyle H}{|}}{C}}-\overset{\overset{\displaystyle H}{|}}{\underset{\underset{\displaystyle H}{|}}{C}}\right]_{n+1}$$

From time to time two of the growing chains will meet and the
one unshared electron at the end of each will be shared leading
to a terminated chain or the final product, polyethylene.

$$R-\left[\overset{\overset{\displaystyle H}{|}}{\underset{\underset{\displaystyle H}{|}}{C}}-\overset{\overset{\displaystyle H}{|}}{\underset{\underset{\displaystyle H}{|}}{C}}\right]_{m}\!\!\cdot \;+\; R-\left[\overset{\overset{\displaystyle H}{|}}{\underset{\underset{\displaystyle H}{|}}{C}}-\overset{\overset{\displaystyle H}{|}}{\underset{\underset{\displaystyle H}{|}}{C}}\right]_{p}\!\!\cdot \longrightarrow R-\left[\overset{\overset{\displaystyle H}{|}}{\underset{\underset{\displaystyle H}{|}}{C}}-\overset{\overset{\displaystyle H}{|}}{\underset{\underset{\displaystyle H}{|}}{C}}\right]_{m+p}\!\!\!-R$$

This free radical chain reaction is the easiest to visualize
and is the process commonly used for the polymerization of the
unsaturated hydrocarbons including the vinyl

$$\overset{\overset{\displaystyle H}{|}}{\underset{\underset{\displaystyle H}{|}}{C}}=\overset{\overset{\displaystyle H}{|}}{\underset{\underset{\displaystyle X}{|}}{C}} \qquad \text{and vinylidine} \qquad \overset{\overset{\displaystyle H}{|}}{\underset{\underset{\displaystyle H}{|}}{C}}=\overset{\overset{\displaystyle X}{|}}{\underset{\underset{\displaystyle Y}{|}}{C}}$$

type monomers. A partial list of the monomers and the polymers
resulting follows. R represents an end group differing from
the repeating unit and usually arises from the initiator but
may be hydrogen, a monomer radical, or another group.

Monomer   Polymer

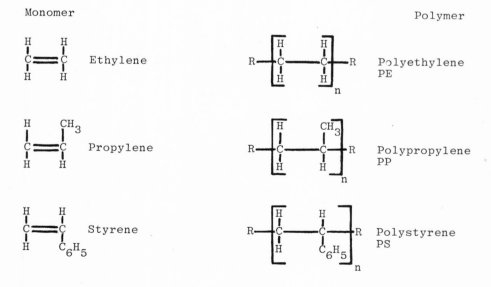

Monomer		Polymer
Ethylene		Polyethylene PE
Propylene		Polypropylene PP
Styrene		Polystyrene PS

Monomer   Polymer

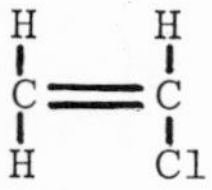

$$\begin{array}{cc} H & H \\ C = C \\ H & Cl \end{array}$$ Vinyl Chloride $$R-\begin{bmatrix} H & H \\ C & C \\ H & Cl \end{bmatrix}_n R$$ Poly(vinyl Chloride) PVC

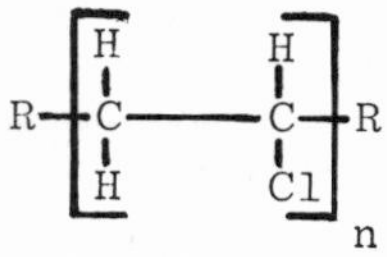

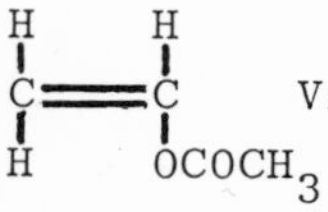

$$\begin{array}{cc} H & H \\ C = C \\ H & OCOCH_3 \end{array}$$ Vinyl Acetate $$R-\begin{bmatrix} H & H \\ C & C \\ H & O \\ & CO \\ & CH_3 \end{bmatrix}_n R$$ Poly(vinyl Acetate) PVA

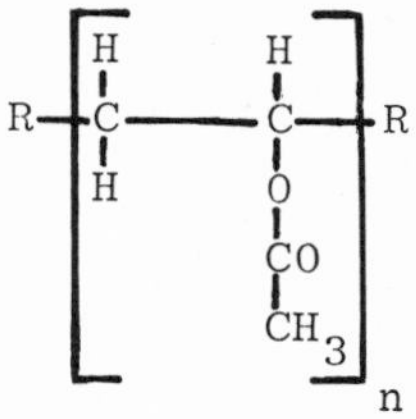

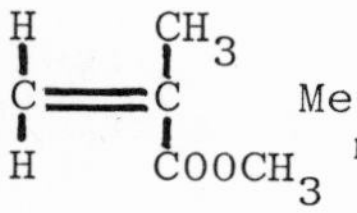

$$\begin{array}{cc} H & CH_3 \\ C = C \\ H & COOCH_3 \end{array}$$ Methyl methacrylate $$R-\begin{bmatrix} H & CH_3 \\ C & C \\ H & CO \\ & O \\ & CH_3 \end{bmatrix}_n R$$ Poly(methyl-methacrylate) PMMA

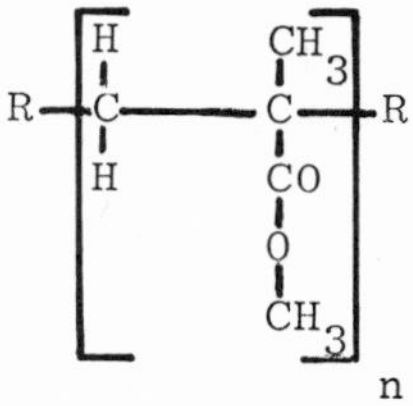

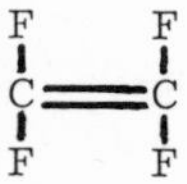

$336 \rightarrow A4$

$A3 \rightarrow A4$

$$\begin{array}{cc} F & F \\ C = C \\ F & F \end{array}$$ Tetrafluoro-ethylene $$R-\begin{bmatrix} F & F \\ C & C \\ F & F \end{bmatrix}_n R$$ Polytetra-fluoroethylene PTFE

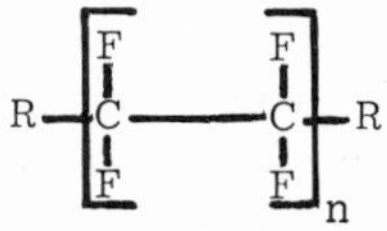

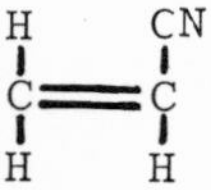

$$\begin{array}{cc} H & CN \\ C = C \\ H & H \end{array}$$ Acrylonitrile $$R-\begin{bmatrix} H & CN \\ C & C \\ H & H \end{bmatrix}_n R$$ Polyacrylonitrile PAN

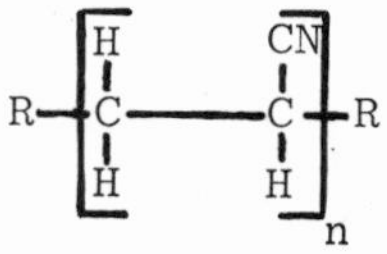

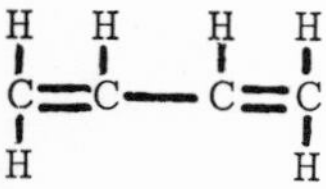

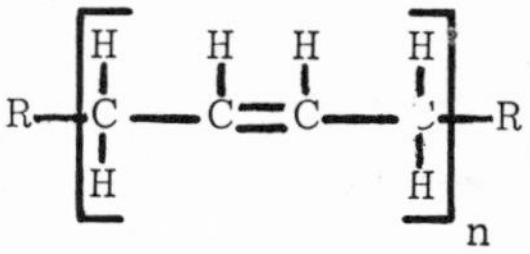

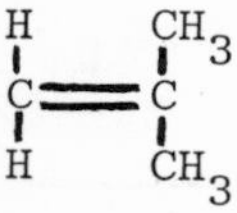

$$\begin{array}{cccc} H & H & H & H \\ C = C & - & C = C \\ H & & & H \end{array}$$ Butadiene $$R-\begin{bmatrix} H & H & H & H \\ C & - & C = C & - & C \\ H & & & H \end{bmatrix}_n R$$ Polybutadiene BR

$$\begin{array}{cc} H & CH_3 \\ C = C \\ H & CH_3 \end{array}$$ Isobutylene $$R-\begin{bmatrix} H & CH_3 \\ C & C \\ H & CH_3 \end{bmatrix}_n R$$ Polyisobutylene PIB

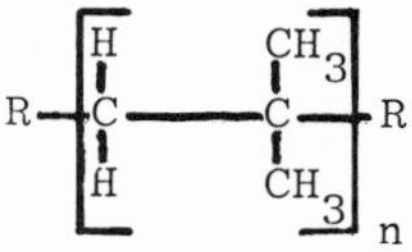

Somewhat similar in general behavior is ring opening polymerization. One of the best examples is the polymerization of ε-caprolactam to polycaprolactam or Nylon 6.

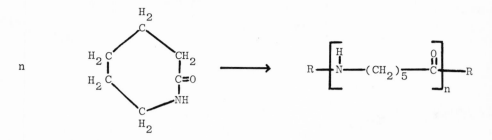

However, in this case the ring of the monomer unit opens and these add one at a time to lengthen the chain. Although an ionic catalyst is usually employed and only the initiated molecules grow the propagation step is more like a chemical condensation and the termination is not a mutual combination of two molecules. Indeed the ring opening polymerization lies somewhat between the free radical addition type and the next type, the condensation or step-wise polymerization. This is illustrated well by the formation of nylon 66 from adipoyl chloride and hexamethylene diamine. In this case the two co-monomers react first in pairs, then more monomers add stepwise with the elimination of hydrogen chloride until the long chain molecule is formed comparatively slowly. This is a conden-sation reaction just like the condensation reactions of small molecules and often there is elimination of some small mole-cule such as water, sodium chloride, hydrogen chloride or methanol.

$$H_2N(CH_2)_6NH_2 \quad + \quad ClC(CH_2)_4CCl \quad \longrightarrow$$

Hexamethylene   Adipoyl
Diamine   Chloride

$$H_2N(CH_2)_6\!-\!N\!-\!C\!-\!(CH_2)_4\!-\!CCl$$

amide

$$H_2N(CH_2)_6\!-\!N\!-\!C\!-\!(CH_2)_4\!CCl \quad + \quad H_2N(CH_2)_6\,NH_2 \quad \longrightarrow$$

$$H_2N(CH_2)_6\!-\!N\!-\!C\!-\!(CH_2)_4\!-\!C\!-\!N\!-\!(CH_2)_6\,NH_2$$

etc.

Nylon 66

Typical monomers and the polymers which result from ring
opening and condensation or step-wise polymerization follow
but it must be realized that a detailed study of the mechanism
of polymerization is necessary to decide which synthetic
process is in fact the important one. Some of the monomers
will polymerization by more than one mechanism.

Monomer(s)   Polymer

Formaldehyde $\longrightarrow$ Polyacetal

3,3'-bis-chloromethyl oxetane $\longrightarrow$ Penton

Ethylene Oxide $\longrightarrow$ Poly-(ethylene Oxide)

2,6-dimethyl phenol $+$ O_2 $\longrightarrow$ PPO

Bisphenol A $+$ $COCl_2$ Phosgene $\longrightarrow$ Polycarbonate

$$CH_3-OC\overset{O}{\underset{}{||}}-\bigcirc-\overset{O}{\underset{}{||}}CO-CH_3 \;+\; HO-\overset{H}{\underset{H}{\overset{|}{\underset{|}{C}}}}-\overset{H}{\underset{H}{\overset{|}{\underset{|}{C}}}}-OH \longrightarrow$$

Dimethyl Terephthalate   Glycol

$$CH_3\left[OC\overset{O}{\underset{}{||}}-\bigcirc-\overset{O}{\underset{}{||}}CO-\overset{H}{\underset{H}{\overset{|}{\underset{|}{C}}}}-\overset{H}{\underset{H}{\overset{|}{\underset{|}{C}}}}\right]_n OH$$

Polyester

$$OCN(CH_2)_6NCO \;+\; HO-\overset{H}{\underset{H}{\overset{|}{\underset{|}{C}}}}-\overset{H}{\underset{H}{\overset{|}{\underset{|}{C}}}}-OH \quad\underset{\longrightarrow}{} \quad R\left[\overset{O}{\underset{}{||}}C-\overset{H}{\underset{}{\overset{|}{N}}}-(CH_2)_6\overset{H}{\underset{}{\overset{|}{N}}}-\overset{O}{\underset{}{||}}C-O(CH_2)_2O\right]_n H$$

Hexamethylene   Glycol   Polyurethane
Diisocyanate

$$HO\bigcirc-\overset{CH_3}{\underset{CH_3}{\overset{|}{\underset{|}{C}}}}-\bigcirc OH \qquad + \qquad Cl\overset{H}{\underset{H}{\overset{|}{\underset{|}{C}}}}-\overset{H}{\underset{}{\overset{|}{C}}}-\overset{H}{\underset{O}{\overset{|}{CH}}} \longrightarrow$$

Bisphenol-A   Epichlorohydrin

$$R\left[O\bigcirc-\overset{CH_3}{\underset{CH_3}{\overset{|}{\underset{|}{C}}}}-\bigcirc O-\overset{H}{\underset{H}{\overset{|}{\underset{|}{C}}}}-\overset{H}{\underset{\overset{|}{O}{\underset{H}{}}}{\overset{|}{C}}}-\overset{H}{\underset{H}{\overset{|}{\underset{|}{C}}}}\right]_n R$$

Epoxy

The preceding polymers were all formed from bifunctional monomers in the sense that when reacted each could combine with two others only whether it was by opening of a double bond or a ring yielding two active ends or by a combination of two chemically active functional groups such as hydroxyl, carboxyl, amino, etc. It is possible, certainly in the last mentioned case, to have three or more functional groups. When this type of monomer is used the polymer is branched and cross linked into a vast, but imperfect three dimensional network. One of the best known plastics made in this way is the phenol-formaldehyde resins in which phenol is in fact trifunctional and the formaldehyde difunctional. Condensation with the elimination

of water leads to a simple, soluble prepolymer and then further
reaction with formaldehyde, usually supplied by urotropin or
hexamethylene tetramine leads to the final network polymer.

These differences in the structure of polymers lead to division
into two main classes, thermoplastics and thermosets. The
thermoplastic polymers may be linear or thread-like, or branched
but regardless of that the molecules are independent and at
some temperature they will be able to flow past one another or
become plastic but when cooled again solidify into a form-
stable material or glassy material with which we are normally
familiar. On the other hand the molecules making up the three
dimensional network polymers cannot flow individually and the
shape is maintained often to temperatures at which the polymer
is decomposed rather than softening or flowing. These are
thermoset polymers or ones set against temperature changes. A
thermoplastic polymer may be made into a thermoset polymer by
putting cross-links between the individual chains. An excellent example of this is the vulcanization of curing of the
raw thermoplastic rubber by sulfur to yield the familiar thermoset vulcanized rubber article.

```
    H       H   H       H
    |       |   |       |
 ——C———C ══ C———C——
    |       |           |
    H       H

         + S8              ———————>

    H       H   H       H
    |       |   |       |
 ——C———C ══ C———C——
    |       |           |
    H       H
```

```
    H       H   H       H
    |       |   |       |
 ——C———C ══ C———C——
    |                   |
    S                   H
    |
    S
    |
    H       H   H       H
    |       |   |       |
 ——C———C ══ C———C——
    |                   |
    H                   H
```

Two units of elastomer chains   Cross linked, cured,
                                        vulcanized

Other cured or thermoset polymers which are in everyday use
are the polyesters used in fiberglass reinforced polyester
products, the polyurethanes, the silicones, the epoxies and
phenoxies, the melamine-formaldehyde and urea-formaldehyde
resins, as well as the phenol-formaldehyde resins mentioned
earlier. The thermoplastic polymers most often cured to thermoset polymers are copolymers of butadiene and styrene, butadiene and acrylonitrile, isobutylene and isoprene, ethylene,
propylene and a diolefine, and ethylene and acrylic acid and
homopolymers of isoprene, butadiene, 2-chlorobutadiene and
ethylene. Extensive monographs may be consulted for details
relating to the individual polymers mentioned above.

Mechanism and Kinetics of Polymerization

It will be obvious that, with such a wide variety of structures and polymerization reactions, there will be a very extensive and complex literature on mechanisms and kinetics of
polymerization [1,2,4]. Only a summary can be included here.
Some polymerization reactions proceed spontaneously or thermally (heat or radiation) but usually an initiator is used as
described for polyethylene. In some instances this initiator
is regenerated or is not consumed during the reaction and is
a true catalyst. Unfortunately the term catalyst is also
applied when the initiator is consumed during the reaction.

If the initiator cleaves into two portions, each having one of the electrons from the electron pair of the bond, two free radicals are formed. These are uncharged but readily react to share the lone electron with one of the electrons of a P-bond of an unsaturated compound such as ethylene leaving the second electron of that bond to repeat the process. This continues rapidly until two such growing chains terminate each other.

$$R : R' \longrightarrow R^O + {}^O R'$$

Free Radicals

$$R^O + n \begin{array}{c} H \quad H \\ | \quad | \\ C = C \\ | \quad | \\ H \quad H \end{array} \longrightarrow R \begin{bmatrix} H \quad H \\ | \quad | \\ C - C \\ | \quad | \\ H \quad H \end{bmatrix}_n$$

Ethylene   Polyethylene

On the other hand the two electrons may go with one part of the molecule yielding an ion pair. Depending upon which ion becomes part of the growing polymer chain the initiator may be cationic or anionic.

$$R : R' \longrightarrow R^{\oplus} + {}^{\ominus}:R' \text{ ions}$$

For example an initiator composed of aluminum chloride and water

$$AlCl_3 + H_2O \longrightarrow H^{\oplus} {}^{\ominus}AlCl_3OH$$

$$C_4H_8 + H^{\oplus}{}^{\ominus}AlCl_3OH \longrightarrow C_4H_9^{\oplus} {}^{\ominus}AlCl_3OH$$

will transfer a proton to isobutene to initiate the polymerization of isobutene which will propagate by the repeated insertion of monomer units between the carbonium ion and the hydroxy aluminum trichloride counter ion.

$$C_4H_9^{\oplus} {}^{\ominus}AlCl_3OH + n\ C_4H_8 \longrightarrow C_4H_9(C_4H_8)_n^{\oplus} {}^{\ominus}AlCl_3OH$$

Termination is by "recapture" of a proton from the end of the growing chain by the counter ion leaving a double bond on the polymer chain. In this way the cationic catalysts may be true catalysts, but this is not a generalization.

The other major type of initiator is the anionic which is illustrated well by lithium butyl. The incoming monomer units,

say isoprene, insert between the Li and the butylene anion and
hence the growing polymer chain is anionic. In this specific
instance there is no spontaneous termination so growth
continues until the monomer is all reacted. The ion pair
remains active and the addition of further monomer, or a different monomer if desired, will lead to further growth and the
formation of a block copolymer in the latter case. The structure of the product depends on the solvent used and is high
cis-polyisoprene when the solvent is hydrocarbon, high 3, 4 -
polyisoprene if the solvent is oxygenated, i.e. tetrahydrofuran.

$$Li\ C_4H_9 \longrightarrow Li^{\oplus}\ {}^{\ominus}C_4H_9$$

Lithium Butyl   Butadiene   Polybutadiene

$$Li^{\oplus\ominus}C_4H_9\ +\ n\ \ CH_2=CH-CH=CH_2 \longrightarrow Li^{\oplus\ominus}\left[CH_2-CH=CH-CH_2\right]_n C_4H_9$$

The reaction can be terminated by water, ethylene oxide or
carbon dioxide to yield hydrogen, hydroxy and carboxy terminated polymers respectively finally.

   While there are many other types of initiators, one which
must be mentioned is the coordination complex catalysts known
familiarly as the Ziegler-Natta catalysts and composed
typically of titanium trichloride and aluminium triethyl
although the variations are numerous. As in the case of the
anionic and cationic initiators the incoming monomers insert
between one of the metal atoms and the alkyl group repeatedly
so that the chain grows out from the initiator. During the
process the configuration of the chain is fixed and this has
lead to the long list of stereoregular polymers and resulted
in these initiators being designated stereospecific.

$TiCl_3$ $Al(C_2H_5)_3$ Ethylene   Polyethylene

   All of the above types of initiators are specially adapted
to addition polymerization. Only certain monomer molecules

are activated and growth takes place only on the activated units. Thus there is present at any one time some fully grown chains, a very few growing chains and unreacted monomer(s). The molecular weight of the product may not change much with time. Ring opening and condensation polymerization reactions may resemble addition polymerization if acid or alkali is required to initiate polymerization but in general any monomer unit can react and the reaction mixture is a complex mixture of dimers, trimers, etc. which gradually grow in length with time. Thus the monomer disappears early in the reaction, the molecular weight increases slowly with time, and growth will only cease when both ends of every chain bear the same functional group. Reaction time and the stoichiometric ratios are the means whereby the molecular weight is controlled, i.e. a careful control of the stoichiometric ratio of functional groups is necessary.

Conformation and Configuration

To this point one can consider the macromolecules as extremely long and thin molecules mixed together like noodles in a bowl, either free to move when a force is applied or tied together at numerous points along the length of each chain so that when a force is applied the whole mass must move, thermoplastic and thermoset respectively. However, there are two stages of order which must be mentioned. The first is that the long chains may align and lie in the same direction, at least for portions of their length. This orientation is important for certain properties to be discussed later. If the structure is regular enough and the forces between chains great enough the chains will form crystals or crystallites, a subject which will also be discussed in more detail later. When the long chain is able to take up many shapes in space we say that it can take up many conformations [5,6], the polymer has not changed structure but merely has arranged itself in different shapes by rotation of the structure about the C-C bonds, a rotation which we can consider freely possible for most of the properties discussed. On the other hand there are cases in which although the chemical structure is the same and conformations are possible there are certain fixed positions for some of the groups of atoms relative to each other, and these are configurations [1,5,6]. Two main types of configurations should be recognized.

The first is that when a double bond $\text{C}=\text{C}$ exists in the chain the chain may pass through that unit in two ways, cis and trans

cis-polybutadiene   trans-polybutadiene

in the former two H atoms are on one side and in the latter on opposite sides of the double bond which is not capable of free rotation in contrast with the C-C bond.

More difficult to visualize are the structures of a polymer such as polypropylene.

isotactic   syndiotactic

This can have three forms, the one in which the methyl groups follow no sequence, one in which the methyl groups are always on the same side of the chain and one in which they are alternately on opposite sides of the chain. By this we mean that as you walk along the chain and approach a carbon atom you will note that the H, the CH_3 and the continuing chain occur in a certain order. If they are all clockwise or counterclockwise the structure is isotactic. If they are alternately clockwise and counterclockwise it is syndiotactic. If there is no uniform sequence, it is atactic. The concept is much like the

D and L configurations of optically active compounds such as
amino acids and sugars but optical activity in a polymer such
as polypropylene would be difficult to detect. The structures
when there are two or more substituents which differ and can
each contribute a sequence have been studied and named but
need not be considered here.

Likewise, although most of the polymers mentioned have been
homopolymers of just one monomer or copolymers of two monomers
in which the monomers either are randomly present as in butadiene-styrene elastomers or alternating as in epoxy resins and
Nylon 66, there are possible at least two other major arrangements. The one is when a branch is composed of a different
polymer than the main chain as for example ABS plastics which
contain styrene-acrylonitrile branches on polybudadiene.
These are called graft copolymers [2,4,5] since the branch is
like a graft on the main trunk. Then there are block copolymers [1,2,4,5] in which all of each of the monomer units will be
together and joined so that you have something like polybutadiene joined to polystyrene. Various monographs treat block
and graft copolymers and comonomer sequence distributions in
detail for specific examples.

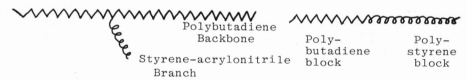

Polybutadiene
Backbone

Styrene-acrylonitrile
Branch

Poly-
butadiene
block

Poly-
styrene
block

Graft Copolymer   Block Copolymer

Molecular Weight Distribution

There is still another factor which must be considered and
that is that a unique molecular weight is not obtained. By
using anionic catalysts, polymers which have very narrow molecular weight distributions may be obtained. However, the
normal distribution of molecular weights can vary from fairly
narrow to extremely broad but may be controlled to some degree
by process variables. As far as most of the studies are
concerned we are dealing with a mixture of molecular weights

and to characterize the polymers we use a viscosity average
molecular weight $\overline{M}_v = \left[\dfrac{\sum N_x\ M_x^{1+\alpha}}{\sum N_x\ M_x}\right]^{1/\alpha}$, a number average molecular

weight $\overline{M}_n = \dfrac{\sum N_x\ M_x}{\sum N_x}$ or a weight average molecular weight

$\overline{M}_w = \dfrac{\sum N_x\ M_x^2}{\sum N_x\ M_x}$. For certain theoretical studies other molecular

weight averages may be calculated but for most practical
purposes the viscosity average molecular weight is used.

The molecular weight distributions [1,2,4,5] for addition
polymers are approximately Gaussian with the number average,
viscosity average and weight average values increasing in that
order. If the product were one unique molecular weight all
three average values would be the same. However, the reverse
is true. Indeed as the distribution of molecular weights
broadens the ratio of the weight average to number average
(M_w/M_n) increases and this ratio is used as a measure of
polydispersity. In addition to the normal distributions one
may have bimodal or multimodal distributions composed of
several overlapping Gaussian distributions, etc.

When condensation types are studied the molecular weight
distribution tends to be skewed towards low molecular weights
as a result of the stepwise formation. A useful equation,
the Carothers equation relates the extent of reaction, the
degree of polymerization (i.e. average molecular weight) and
the average functionability of the monomers used in the system.
Variations of this equation are used for specific monomer
systems. Addition polymers often have distributions skewed
to high molecular weights due to branching, cross linking and
other secondary reactions taking take at higher conversions
of monomers to polymer.

The relationship between kinetics and mechanisms and the
average molecular weights of the products have been studied
for specific cases and may be found in the more exhaustive
texts.

With this very brief introduction to the nature of polymers
the subject of their structure and preparation will not be
discussed in detail further. Rather the properties and perfor-
mance will be treated without reference to specific polymers

in most instances. Readers should recall that various
chemical and structural variations are possible. For example
a long regular chain molecule might well crystallize whereas
a short irregularly shaped chain molecule might remain a
viscous fluid, the former being a useful hard plastic and the
latter a useful sticky adhesive or sealant.

2

Interaction Between Polymer Molecules

Fundamental to many engineering aspects of polymers is interaction between the molecules. For small molecules it is comparatively easy to obtain some measure of this interaction by measuring the latent heat of evaporation, for example. A comparable value can be estimated for polymers and has considerable value in characterizing the interaction between the molecules even though the molecules cannot be evaporated. The cohesive energy density (CED) is defined as the latent heat of evaporation per unit volume, for example, calories per millimeter [5,7,8]. The square root of the CED is defined as the solubility parameter (δ) and has the units of cal. $^{\frac{1}{2}}/\text{ml}^{\frac{1}{2}}$. Numerous tables of values for the CED or the solubility parameter (δ) may be found in texts but a diagram by Small [9] illustrates the range of values well and the tables in the Polymer Handbook are particularly useful [10]. Some typical values are listed in Table 2-1.

The CED of Soluble Polymers

While the values for the CED or δ may be found readily for the solvents, how are they obtained for involatile polymers? There are two general approaches. The first is based on the viscosity of solutions of the polymer in various solvents, at a constant concentration of solute or polymer. When a soluble polymer dissolves the chains extend and take on random conformations. There is a gain in entropy. If the solvent molecules and the polymer molecules have equal or comparable attractions for each other the solvent molecules and the units of the polymer molecule are equally at home and complete randomness can result. The polymer chain will extend to an equilibrium

Table 2-1.

Typical values of the solubility parameter and the H-bonding index selected from the Polymer Handbook [10] and Beerbower et al. [11].

Solvent	Solubility Parameter	H-Bonding Index	Polymer	Solubility Parameter
Dimethyl-siloxane	4.9-5.9	poorly		
Perfluoro-heptane	5.8	poorly	Polytetra-fluoroethylene	6.2
Propane	6.4	0.0		
Heptane-n	7.4	0.0	Poly(dimethyl-siloxane)	7.4
Diethyl ether	7.4	13.0	Polyethylene	7.9
Cyclohexane	8.2	0.0	Polyisobutylene	7.9
Carbon tetra-chloride	8.6	0.0	Polyisoprene	8.1
Diethyl ketone	8.8	moderately	Polybutadiene	8.4
Tetrahydrofuran	9.1	moderately	Polystyrene Polypropylene	8.9 9.2
Benzene	9.2	0.0	Poly(vinyl acetate)	9.4
Chloroform	9.3	1.5	Poly(vinyl chloride)	9.8
Acetone	9.9	9.7	Poly(ethylene-terephthalate)	10.7
1,4-Dioxane	10.0	9.7	Poly(vinylidine chloride)	12.2
Carbon disulfide	10.0	0.0	Polyacrylonitrile	12.6
m-Cresol	10.2	strongly	Poly(hexamethylene-adipamide)	13.6
Dimethyl phthalate	10.7	moderately	Poly(methyl-α-cyanoacrylate)	14.0
Acetonitrile	11.9	poorly		
Dimethyl sulfoxide	12.0	moderately		
Dimethyl formamide	12.1	moderately		
Nitromethane	12.7	2.5		
Methanol	14.5	18.7		
Water	23.4	39.0		

value which will be as random as possible. From other studies
this should result in a maximal viscosity of the solution
taking into account, of course, the viscosity of the solvent.
If the polymer segments attract each other more strongly than
they do solvent or if the solvent molecules attract each other
more strongly than they do polymer, then the polymer chain will
be less extended and the viscosity of the solution will be less.
Thus a curve will result which shows a maximum viscosity for
the solvents which have CED values close to that of the polymer,
or conversely the CED of the polymer will be that of the solvent
corresponding to the peak in the curve.

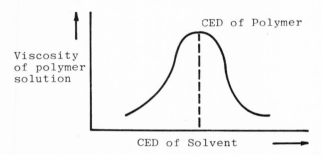

   There are several useful concepts which relate to the above
which fit best into the discussion here. Firstly, the solubility and the intrinsic viscosity will vary with the temperature,
and vary differently for each polymer and solvent. Thus it is
possible by varying the temperature or solvent to achieve a
condition which physically is equivalent to a polymer molecule
of infinite molecular weight being just soluble. This temperature for the particular solvent is the theta (θ) temperature,
or the particular solvent is the (θ) solvent for the polymer in
question. The importance of this will be emphasized when the
viscosities of polymer solutions is discussed later.

   Secondly, mention was made that when the CED of the solvent
and polymer were the same then the polymer was most highly
extended. By various techniques but particularly by light
scattering it is possible to measure a root mean square end-to-
end distance of molecules in solution and the radius of this
"sphere" can be designated $(\overline{r}^2)^{\frac{1}{2}}$. Under other experimental
conditions the root mean square end-to-end distance will be

less and the ratios of these values to the theta value $(\bar{r}_o^2)^{\frac{1}{2}}$
are the expansion factors, α

$$\alpha = \left[\frac{(\bar{r})^2}{(\bar{r}_o)^2}\right]^{\frac{1}{2}}$$

The expansion factor is likely to be a maximum in a solvent
with the same CED as the polymers.

Thirdly, it will be immediately obvious that by mixing a
solvent with a high CED with one with a low CED the CED of
the mixture should equal that of a polymer in the intermediate
range. This is indeed the case but the technique is not free
of problems. If one of the solvents is more strongly attracted
to the polymer molecules or to one specific part of the polymer
molecules such as a side chain than the other solvent then the
solvent medium will not be uniform on a micro scale. The
solvent attracted to the polymer will be more concentrated in
the neighbourhood of the polymer and the bulk of the solvent
will be depleted of the solvent more highly attracted. Nevertheless a judicious choice of solvent may be most useful. In
making polymer solutions for adhesives or cements it is often
useful to mix solvents one of which is good and one of which
is a non-solvent which causes phase separation whereby the
polymer and good solvent are dispersed and the mixture yields
a less viscous cement or solution with the poor solvent the
continuous phase. Carried still further the addition of more
non-solvent will cause the largest polymer molecules to
separate from solution and by repeating this process many
times a fractionation of the polymer with respect to molecular
weight can be achieved, a technique described in many texts on
polymer characterization [5,7].

Since it can be presumed that the attraction between molecules will be related to the functional groups present, Small[9]
listed the molar attraction constants for typical groups and
calculated the solubility parameters with astonishingly good
results. A few of his data are in Table 2-2. Note that his
calculations spanned the range from polyisobutylene ($\delta = 8.1$)
to poly(vinyl chloride) ($\delta = 9.6$) at least.

Table 2-2.

Calculated and observed solubility parameters for typical polymers.

Polymer	δ calc.	δ obs.
Polyisobutylene	7.7	8.1
Polyethylene	8.1	7.9
Polyisoprene	8.2	7.9 - 8.4
Polybutadiene	8.4	8.4 - 8.6
Polystyrene	9.1	9.1 - 9.2
Poly(methylmethacrylate)	9.3	9.3
Poly(vinyl chloride)	9.6	9.6

Furthermore, Small suggested the proton attracting power might be a modifying factor and thus applied a plot of the iodine bonding number (I/mole) versus the solubility parameter to yield a map of the solvents and non-solvents for poly(vinyl chloride) (solubility parameter 9.5). An adaptation of his diagram is in Figure 2-1.

Detailed discussion of the data is in the original manuscript. It will suffice here to indicate that the solubility parameter of the solvent alone is not good enough to define exactly the δ of the polymer. The difficulties become even greater when the polymer is insoluble in most if not all solvents. For example cross linked fluorocarbon elastomers merely swell in most solvents and a map somewhat similar to that of Small was prepared by Beerbower, Kaye and Pattison[11] in which the hydrogen bonding index (see Table 2-1) of the solvent was plotted against the solubility parameter and lines of equal swelling of a cross linked fluorocarbon elastomer drawn. An adaptation of these data is in Figure 2-2. The polymer is VITON, a copolymer of vinylidine difluoride and perfluoropropene. The δ as measured varies from 6.6 to 8.3, considerably below what the data from the swelling measurements of the crosslinked polymer would indicate.

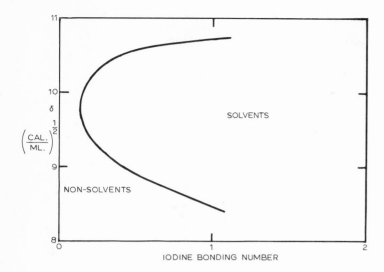

Fig. 2-1. Iodine Bonding Index versus solubility parameter for various solvents and non solvents for poly(vinyl chloride). (After Small [9]).

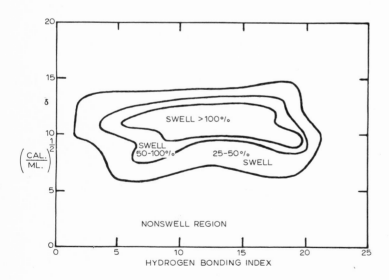

Fig. 2-2. Swelling of cross-linked fluocarbon rubber as a function of the solubility parameter and hydrogen bonding index. (After Beerbower et al. [11]).

The CED of Cross Linked Polymers

The vinylidine difluoride-perfluoropropene copolymer may be atypical. The above variation of the classical technique for measurement of the CED of cross-linked or insoluble polymers has not been used as much as the original method which was simply to lightly, but uniformly, cross link the elastomer (or take an already cross linked product) and swell samples of the product in solvents of differing CED values [12] The maximal

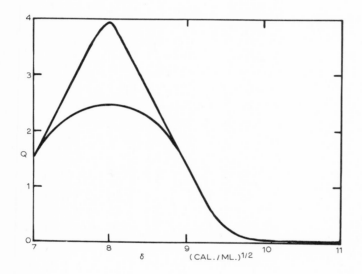

Fig. 2-3. Swelling of natural rubber in various solvents: (Gee [12]). upper, pure gum; lower, tire tread vulcanizate containing 60 parts of non-rubber compounds per 100 parts of rubber rather than 10 parts for the gum compound. Swelling was calculated on the basis of the actual rubber content.

swelling takes place in the solvent with the same CED as the polymer for the same reason that the viscosity of the solution was greatest, the chains of the polymer and solvent attain their most random conformations. Addition of a solvent which differs in solubility parameter from that of the polymer would result in deswelling to a new equilibrium. The swelling of

the gel or insoluble polymer is balanced by the stretching and
elongation of the chains in the network. The solvent has the
same activity as in a solution; the partial molar free energy
of the solvent in a gel is the same as for the pure solvent.
Arising from this, one can calculate an interaction parameter
X_1 which will be discussed in detail with all its implications
in texts [7] particularly relating to polymer solutions, and of
course it is useful in studies of the swelling and deswelling
of gels or swollen polymer in contact with solvents and solutions.

Still another complicating feature is that crystalline
regions in polymers may not be penetrated by solvent until
near the melting point, hence spurious results may be obtained
unless the polymer-solvent mixture is heated. A practical
application of this principle will be described later under
the topic of plastisols.

Another principle to keep in mind when considering polymers
is mutual incompatibility of polymers. This is almost a
universal rule but there are a few exceptions. The sensitivity
of the technique is admirably illustrated by the two phase
system, natural rubber and polystyrene in benzene published
by Dobry and Boyer-Kawenoki [13]. Note that a little over one
percent of either natural rubber or polystyrene will cause
the two phases to form, containing not over 1% of either of
the polymers. A solution of about 5% SBR (styrene-butadiene
rubber), or BR (polybutadiene rubber) and 5% polystyrene
in benzene, toluene, ethyl benzene or styrene will separate
into two layers rapidly. The volume of each layer will depend
upon the relative amounts of the two polymers and separation
will cease when the viscosity of the mixed solutions becomes
too great for separation by mass movement under gravity. It
will be noted that the solubility parameters of the two polymers and the solvents do not differ greatly, 8.1 to 9.1. The
data of Dobry and Boyer-Kawenoki reproduced in Figure 2-4 are
plotted in a far from conventional ternary diagram form but are
easier to interpret when such a small corner of the complete
diagram is used. However, it is easy to see that as the percent of either rubber or polystyrene increases the separation
of the polymers into two separate solutions becomes almost
complete.

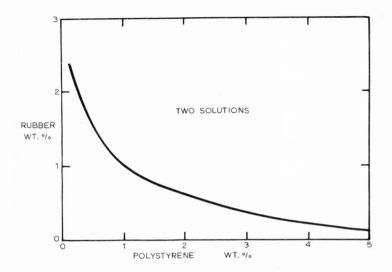

Fig. 2-4. The ternary system benzene-rubber-polystyrene in the dilute region. Dobry and Boyer Kawenoki [13].

Calculations of the CED

As a final comment on the CED one can quote two attempts to calculate the values based on the physical state of the polymers used, i.e. the glass transition temperature. This temperature will be defined and discussed in detail later but for now may be considered the temperature at which a molten polymer (polymer melt) turns to a glass upon cooling. Polystyrene has a glass transition temperature well above room temperature, about 80 to $100^{O}C$, whereas cis-polybutadiene has a glass transition temperature well below room temperature, perhaps -100 to $-110^{O}C$.

Hayes [14] obtained a linear relationship between the glass transition temperature (absolute scale) and the cohesive energy density divided by a number (n) analogous to the degrees of freedom in expressions of kinetic energy.

Wolstenholme [15] used a more complex function of the cohesive energy E_c, the repeating unit molecular weight, $\overline{M}$, the density of the polymer, ρ, and the length of the repeating unit in angstroms, L. Their relationships are illustrated in Figure 2-5. The fact that the lines are parallel is interesting but not unexpected since the same data are involved multiplied by a different estimate of the polymer dimensions.

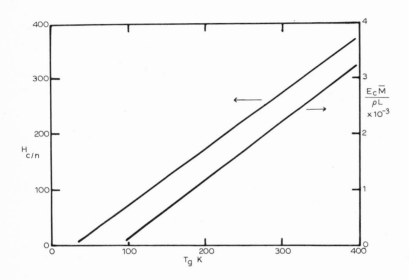

Fig. 2-5. Cohesive energy density versus glass transition temperature as plotted by Hayes [14] left hand side and Wolstenholme [15] right hand side.

Gels and Swollen Polymers

By this time one might well ask what value is the CED or the solubility parameter. Naturally the value is greatest when only one polymer is considered and of course it is a measure of a useful value not otherwise obtained easily. The fact that corrections may need to be applied and more complex systems

misbehave, that crystallinity of the polymer interferes, and
that incompatibility of polymers, perhaps due to differences
in the solid state properties of the polymers (i.e. whether
glassy or a melt at the test temperature) represent warnings
that when polymers are used in concentrated solutions, deviations might be expected which can be accounted for. Some of
the first observations of a practical nature which one observes
are related to the above principles.

For example, polymer solutions or swollen gels are often
very tacky or sticky, a fact related to the mobility of the
chain segments which are able to reach and penetrate the
adherend when they are used as adhesives. Likewise the long
chain molecules may be so entangled that the solution or
swollen gel holds together and can be extruded or spun into a
thread from which the solvent can be removed before the molecules slip past one another. Parenthetically it can be noted
that as a solution of a polymer is spun or caused to flow
through a tube the molecules tend to align in the direction of
flow giving some orientation in the resultant dried thread,
and some observable optical activity, streaming birefringence,
to the solution in motion. Finally the swollen molecules or
gel may be used as cements, thickened fuels, self sealing
insulation for electrical and telephone cables, and flame-
thrower fuels.

The topic of solution extrusion or spinning of fibers
mentioned above is being dismissed too briefly since there are
many interesting aspects to the preparation of solutions, the
adjustment of viscosity for optimal spinning, and the removal
of solvent by precipitation of the polymer with non-solvent or
by evaporation. However, this subject more properly belongs in
a text on textile science. Rather, to illustrate the importance
of the CED and solubility parameter in a field of more direct
interest to the polymer engineer one can use plasticized polymers which unlike the polymer solutions about which we have
been writing, usually are composed of less plasticizer than
polymer. Typically the plasticizer content varies from 5 to 70
parts of polymer depending upon the use and the properties
desired.

Plasticizers and Plastisols

Plasticizers are non-volatile solvents the purpose of which is to make the mixture of polymer and plasticizer softer, flexible, and more easily worked. Normally the rigid polymer molecules are held together by van der Waals forces, hydrogen bonding, polar forces and, as described earlier, these bonds may be broken and the molecules separated by a solvent. This process may require heat. There may be crystalline regions into which the plasticizer (solvent) penetrates with great difficulty and non-crystalline regions into which it penetrates slowly without heating. Thus rigid poly(vinyl chloride), for example, may be quite insoluble at moderate temperatures in solvents or plasticizers having the same CED as the polymer so that a dispersion of the polymer in the plasticizer may be made which is essentially a stable non-aqueous dispersion. The dispersion is called a plastisol and it may have the consistency of slush.

When this plastisol is heated for a few minutes at about 150°C the plasticizer penetrates the polymer and they combine into a uniform mass which is now soft and flexible, the hard, rigid, plastic poly(vinyl chloride) has been plasticized by the non-volatile solvent or plasticizer to yield the flexible, vinyl materials with which we are familiar such as vinyl coated fabrics used for boots. The mixture can be molded and fabricated in many ways. One of the processes is called slush molding and another rotomolding. Later we shall describe what the plasticizer has done but at this stage it will suffice to say that the plasticizer molecules are dispersed between the polymer molecules, poly(vinyl chloride) in this case, so that these molecules are held apart and are able to move more easily. This is noted as a greater free volume and a lower glass transition temperature.

The use of plasticizers of course need not be by the plastisol process only and indeed the usual process is merely to mix the plasticizer(s) and polymer(s) during the compounding or mixing steps at elevated temperatures to achieve the desired product having the flexibility and strength required. This subject will recur again from time to time but one must recall

that a compatible or good plasticizer will have a CED very
close to that of the polymer whereas other organic materials
which differ markedly from the polymer as far as the CED is
concerned will not act as a plasticizer, not make a plastisol,
and may separate readily or bleed from the mixture with time
forming a second phase or a surface film respectively.

The use of volatile solvents yields what is known as an
organosol but the process is exactly the same except that the
volatile solvent leaves the polymer behind in its original
rigid form, like a lacquer for example. Also a polymerizable
(monomer) solvent may be used and after the plastisol has been
made uniform this monomer is polymerized to yield a plastic,
the two interpenetrating networks of polymer molecules forming
a rigidsol. Sometimes to plastisols, organosols or the rigid-
sols one adds a thickening agent before use so that the mass
is thixotropic or more easily spread onto a surface such as a
fabric in a thick layer without running. The whole subject is
very broad and only the main concepts have been touched on.

Plasticizers are usually present in amounts from 5 to 30
parts per 100 parts of polymer. When 30 to 70 parts are used
the term plasticizer is still used but extension is often
implied, more plasticizer is added than is needed to soften the
polymer and the mixture is hardened again by adding a cheap
filler such as carbon black, clay, talc, magnesia, or calcium
carbonate. If still more, say 70 to 500 parts is added the
products are heavily loaded with fillers and are called mastics
or caulks.

If the amount of plasticizer exceeds that which can be
retained by the polymer or the polymer and filler the excess
will come to the surface, bleed, and can be felt or in extreme
cases will actually separate. Plasticizers in addition to
imparting softness and flexibility, may of course, result in
cheaper compounds for certain purposes and may, by being rich
in chlorine or bromine, impart flame retardency to the mixture.
This subject will be discussed later.

While not nearly as well known or as often used as plasticizers there are antiplasticizers. These stiffen polymers,
lower the free volume, and increase the glass transition temperatures but like plasticizers they are non-volatile solvents. The

classical examples are chlorinated aromatics in polycarbonate resin which has a large free volume. The chlorinated wax, "arochlor", fits between the chains but holds them together so that they cannot separate or move freely, just the opposite to the traditional effect of a plasticizer which allows movement.

To polymer fabricators the compatibility of small organic molecules with polymer molecules as indicated by the CED or the solubility parameter has many uses and is very important whether it be in surface coatings, adhesives, plastisols, spinning of fibers, sealants and caulks, or mastics. A knowledge of the approximate values for the δ of the polymers and the various solvents and plasticizers used in a practical system may well explain production and end-use problems not attributable to the quality of any individual ingredient in the formulation.

3

The Amorphous State

Macromolecules or polymers have been described firstly as long thread-like molecules which may be branched or cross-linked into networks. Then we noted that the molecules interacted with each other and that this interaction was measured indirectly by studying the effect of solvents on the polymers. The polymer solutions had many uses beyond being just a method of characterization. Non-volatile solvents are used to make the polymeric masses more flexible and softer, for example. Now let us turn to a study of the polymeric mass as a system whether there is plasticizer present or not and discuss the properties of the mass first as to its structure and behaviour under static conditions, then under dynamic conditions. Later we shall study some practical use conditions, and, finally, deterioration under physical stress, and corrosive and harmful environments.

Structure in the Amorphous State

Polymers in the molten state can be visualized as a bowl of spaghetti or perhaps more correctly as a can of worms since the molecules do not lie still but are under constant thermal motion. Except for their giant size these molecules can be considered just like smaller molecules which make up gases, liquids and solids with which we are familiar. The difference is that in polymers the "small molecules" become joined together and thus there are constraints on their movement. Nevertheless as for all matter there is a certain amount of free volume between the molecules which leads to the possibility of compressing the mass slightly, its compressibility. It also allows localized movement of portions of the molecules.

If the center of gravity of the whole molecule moves, this is translational motion and usually is associated with the application of an external force. If only segments or portions of the molecule can move the center of gravity generally will remain fixed and we have just segmental motion. Both types of motion take place in liquids, neither in solid glassy polymers, and segmental motion only in flexible elastomeric or leathery polymers. We soon notice that these effects are very sensitive to temperature and pressure and also that there are two macroscopically obvious structural conditions or morphologies, one which is called amorphous or isotropic in which the properties are the same in all directions, and another which is anisotropic or in which the properties depend upon the direction of the measurement, a condition which we shall later study as orientation and crystallization.

The isotropic or amorphous condition is easier to visualize and we shall study it first. The assumption made above was that it has the same properties in all directions like glass or a liquid at rest. For most measurements and on a gross scale this is true. The real interest in this case lies in the effect of temperature on the properties, to which we shall return later. However, on closer examination we do find some semblance of order or incipient order even in the amorphous state. There is evidence that the molecules of polytetrafluoroethylene are like balls of string. Other molecules tend to clump with their ends together similarly to polar liquids and liquid crystals. There are orientations and conformations which involve only small portions of molecules or a few molecules and the oriented portions are in turn disordered so that the over all effect is one of an isotropic or amorphous material.

Collier [16] has summarized the main conditions. The "gas-crystalline" theory can be invoked to explain some of the local order in the molten amorphous state. The gas-crystalline theory implies that there is a two dimensional lattice which only requires orientational melting for complete disorder. This idea of ribbons of oriented polymer is not unrealistic. It is akin to the nematic state of smaller molecules. Also when poly(ethylene terephthalate) is melted a globular, paracrystalline, mesophase, liquid crystal type of order is retained any may be observed by birefringence measurements. A smectic state has been observed by electron micrographic studies of such systems. Furthermore polymer crystals have a surprising tendency to reappear in the same positions and in the same numbers upon cooling of a melt made from a crystalline or semicrystalline polymer. If this sort of order persists from the crystalline end of the scale it can be inferred that there is an incipient formation of this type of order from the amorphous end of the scale, a continuous change from one to the other. Diagrammatically nematic and smectic states may be illustrated as follows:

NEMATIC   SMECTIC

Studies of morphology on a gross scale also suggest that the molecules may be organized in at least two extreme ways and presumably intermediate cases. Globular arrangements would result if the polymer molecules were rolled up like a ball of string with interaction only within the globules as was suggested earlier for poly(tetrafluoroethylene). On the other hand some of the molecules may align by interaction along the chain to form fibrillar bundles. Calculations made of the expected conformations of single molecules often yield data which suggest that the molecules are more extended or stiffer than one would expect.

To close this discussion of order in a disordered state we can list various driving forces which would tend to bring about some order, slight though it might be. These are the intermolecular forces variously called van der Waals, London dispersion, dipole-dipole, H-bonding, ionic and electrostatic. Long side chains and other structural aspects of the individual molecules may assist. Cross-linking, branching, very high molecular weight and viscosity, the presence of fillers and

reinforcing agents may all constrain the molecules so that
they form ordered regions. Heating, quenching and annealing
may favor some order. The application of stress, the presence
of high surface energy, and the softening effect of solvents
or plasticizers, may allow the necessary movement. So, order
in the amorphous state though not a characteristic of that
state on a gross scale may exist in small domains and may well
explain some of the anomalous properties of amorphous polymers.
In some recent literature the term "amorph" has been suggested
for these coherent or loosely ordered domains.

Glass Transition Temperature

While the morphology or detailed structure of amorphous or
isotropic polymers may be difficult to observe and not well
known or appreciated, there is no doubt about the changes which
take place as the polymer is heated or cooled. The first such
change is one which continues to involve only the amorphous
state. If a polymer is molten and cooled there will be a
temperature at which it changes from an amorphous, viscous
liquid to an amorphous glass, i.e. there will be a rapid
increase in viscosity to a very high value so that for most
purposes we can consider the polymer a solid glass. It need
not be transparent but it is so for such polymers as polystyrene and poly(methyl methacrylate). To give insight into
what is taking place it is convenient to follow the specific
volume as a function of temperature.

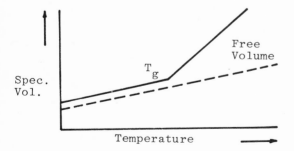

The dashed line is what one would expect taking into account
the normal free volume of molecular systems. In practice the

free volume of polymers is higher by about 2.5%. Of more
importance is the observation that at the temperature at which
the glass softens to a liquid or conversely the liquid solidifies to a glass there is a change in slope of the measured
values, a second-order transition, and the free volume
increases more rapidly with temperature, reaching values as
high as 16% for some polymers. This increased free volume
over what one would expect is sometimes called the WLF free
volume [7] after the authors Williams, Landel and Ferry. The
temperature at which this change in slope takes place is the
glass transition temperature. Some typical values are in
Table 3-1 [10].

Table 3-1

Representative glass transition temperatures

Polymer	Glass Transition Temperature $^\circ$C
Poly(dimethyl siloxane)	-123
Cis-polybutadiene	-108
Polyethylene, amorphous	-78
Poly(vinylidine chloride)	- 19
Nylon 66	57
Poly(ethylene terephthalate)	69
Poly(vinyl chloride)	81
Polystyrene	100
Poly(methyl methacrylate)	105

While the glass transition may occur at any temperature, for
most of the useful thermoplastics it lies between 320 and 420 K
if the polymer is amorphous. If it is semicrystalline, as will
be described later, the glass transition temperature of the
non-crystalline regions is between 170 and 270 K for the well
known types.

Again, although the change in slope is usually shown
diagrammatically as sharp, there is in fact curvature and the
location of the curved portion is a function of the technique
of measurement. The attainment of equilibrium in such viscous

systems is slow so that the final volume, or equilibrium
volume, is attained only after considerable time and this
shrinkage, which eliminates voids or extra free volume, may be
speeded by allowing the diffusion of the chain segments, i.e.
viscous flow, to take place by keeping the temperature at as
high a value as possible, just above or at the transition
region and by avoiding rapid quenching of the melt to a glass
which will "freeze in" free volume which will only disappear
over a long period of time as shrinkage. The changes can be
shown diagrammatically.

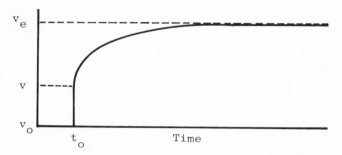

where t_o and v_o are the initial times and volumes, v is the
easily accomplished change in volume and v_e is the final
equilibrium volume.

Transitions and Relaxations

Before describing what has happened on a molecular scale
when a glass changes to a melt and vice versa we might consider
a diagram from Boyer's reviews [17] which shows the effects of
temperature and of test frequency.

Fistly the vertical axes represent energy absorptions or
responses to sinusoidally applied forces, mechanical for the
top curve, electrical for the center curve, and electromagnetic
for the bottom curve. The frequencies increase in the order 1,
to 10^3 to 10^7 Hz. The peaks which represent corresponding
changes in the polymer system move to higher temperatures with
increasing frequency. Secondly there are several peaks or
changes noted. These are discussed in additional detail by
Boyer [18,19,20] and in the monograph by Haward [21].

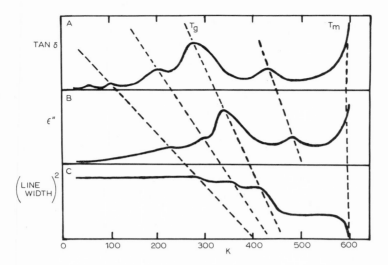

Fig. 3-1. Schematic representation of the energy absorption of
a partially crystalline polymer by three techniques: A, dynamic
mechanical absorption tan δ; B, dielectric energy absorption
ε'; and C, nuclear magnetic resonance. The numerical values
for the vertical scales may be ignored for this representation.
Curves B and C may be continued past the T_m in practice.

Of the peaks shown that marked T_g is the glass transition
temperature, the β-transition, a second order transition. At
the extreme right is T_m the melting point or α-transition, a
first order transition, which appears little affected.
Crystallinity and the melting of crystals will be discussed
later. There are several other peaks noted. One between T_g
and T_m is often called $T > T_g$ and there is one below T_g often
designated $T < T_g$. There are others at still lower temperatures, designated γ-relaxations.

   The change which takes place at the glassy transition
temperature upon heating is that units of polymer chains of
perhaps 50 to 100 atoms begin to move. The number of atoms
is arbitrary and some authors prefer a lower number, say 20
to 50. The atoms need not be and indeed probably would not be

in the same molecule. It is sufficient that at that temperature they are able to move. It is a philosophical problem whether the increase in free volume allows them to move or whether the decrease in viscosity upon heating allows them to move creating the increased free volume. For the engineer it makes no difference. At T_g the amorphous glass becomes an amorphous viscous melt upon heating. In physical terms this means that the polymer molecules which have been immobile are able to move at a rate comparable to the rate of heating and hence respond to an applied disturbance corresponding to the average relaxation rate of the polymer molecules. This rate has increased with temperature until it has reached, near T_g, the required value. The movement and relaxation results in energy absorption which increases rapidly as T_g is approached. Above T_g (or T_β) the molecules move still more freely, the energy loss is less and one gets a peak in energy absorption near the glass transition temperature, the β-loss peak or peak of energy absorption for the β-relaxation. The energy absorbed and released on relaxation appears as heat, an important factor for discussion in later chapters. This explanation also gives one at least a crude explanation of the observation that rapid rates of heating result in a higher measured values of T_g.

The relaxation $T > T_g$ is quite reproducible and not at all easy to explain. It appears in only a few cases and it has been suggested that it represents the motions of still larger segments or whole molecules perhaps to some partially oriented state such as was discussed earlier. The peak labelled $T < T_g$ is one of those which may be caused by a crankshaft rotation of four to eight atoms of the chain. These relaxations as well as relaxations which occur at still lower temperatures in general, the γ-relaxations are not transitions comparable to T_m or T_g. They are suspected to have a role in the energy absorption and the impact strength of polymers since, although the energy absorbed by each unit may be small, there are so many of them. These γ-relaxations are attributed to rotations, oscillations or vibrations of smaller groups such as methyl, methylene, phenyl, phenylene, carbonate, amide and ester present in the chain backbone or as side groups on the chain

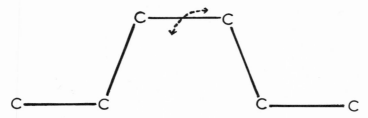

Simplest Crankshaft Motion

backbone which when disturbed relax back to equilibrium positions at characteristic rates releasing to the surrounding
medium the energy absorbed.

   Whether one attributes the glass transition temperature to
a viscous or mass effect, or to a configurational, conforma-
tional, orientational, stiffness or geometric effect is con-
troversial and need not affect the utility of the concept. It
should be noted that mention was made earlier that the glass
transition temperature could be related to the cohesive energy
density and the cohesive energy density obviously will effect
both viscosity and orientation.

   In summary the glass transition temperature and β-relaxation temperature and to some extent the other relaxation temperatures but not the melting temperature depend upon the
existence of free volume for movement and energy for disturbing
the moving unit. The T_g is measured at a higher value at
increased rates of heating, at increased pressures, when the
polymer is branched or cross linked, and at higher test frequencies all of which would tend to either reduce the free
volume or reduce the time for the molecules to relax to their
new condition. On the other hand the T_g is measured at lower
temperatures the slower the rate of cooling and in the presence
of a good plasticizer. The former allows more time for the
polymer segments to flow into their new equilibrium conditions
and the latter promotes the flow of the polymer segments by
reducing the viscosity and increasing the free volume. Other
factors can affect the results. Some of these will be mentioned later, as will be the importance of the various transitions in practice.

Applications of the Glass Transition Concept

The importance of the time factor can be illustrated by quoting an example based on polystyrene [22]. If the polymer is heated to 500 K it is a viscous liquid and may be extruded in the form of a rod into a cooling bath at 300 K, i.e. quenched into a glassy solid. If the pieces to be cut from the rod must not shrink more than 1% after being cut when can you cut them and still meet this specification? The true specific volume is 0.9575 ml/g. The quenched sample has a specific volume of 0.9825 ml./g. and after 30 hours it would change to 0.9672 ml./g. i.e. to within 1% of the specification value. If on the other hand the rod were held at 350 K which is just below T_g for as short a time as 2 hours the shrinkage would take place and the rod would be within specification limits. Heating below T_g is of course necessary otherwise the rod would flow and lose its shape. These are similar problems with semicrystalline polymers such as nylon and polyethylene in which shrinkage may occur due to continued crystallization. The former when first molded may be too large and some weeks may elapse before nylon parts will pass inspection. In the latter case recycled polyethylene milk jugs gradually shrink and after a number of cycles, perhaps 12, must be ground and reblown so that they will hold the legal volume of milk. Parenthetically one might say that making the bottles too large to start with might make them pass through the plant more often but the automatic filling machines would leave an air gap at the top during the early cycles and the customer would feel cheated.

At the risk of attributing too much significance to T_g and the relaxations at lower temperatures it is convenient to discuss here two practical problems which may be related to the glass transition in part. Again we must employ the energy absorption or energy loss factor, tan δ, which will not be defined precisely until later. It must suffice for now to say that when this increases to a peak the polymer will absorb most energy at that temperature or test frequency.

The first is the process of making high impact polystyrene[17] By the addition of about 6% of an elastomer, usually

polybutadiene, to polystyrene during its manufacture the
product, a polyblend, is considerably more resistant to impact
though still not outstanding so. It will be noted, Figure 3-2,
that the presence of the rubber in the material not only moves

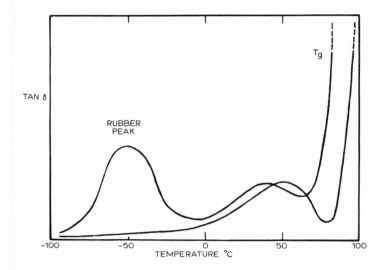

Fig. 3-2. Schematic dynamic mechanical loss curves for regular
polystyrene and rubber modified polystyrene. Note that the
tan δ peaks corresponding to T_g and $T < T_g$ for polystyrene
move to lower temperatures and that the tan δ peak corresponding to the T_g of the rubber appears.

the T_g and the $T < T_g$ peaks to slightly lower temperature but
a peak attributable to the rubber appears. The presence of
the rubber is not sufficient to account for the absorption of
energy on impact and the usually accepted theory is that the
rubber exists in small domains about which minute energy
absorbing cracks or crazes form which heal again with time.
This subject also will be described under failure phenomena.
In the meantime it is tempting to suggest that the lower β-
relaxation temperature contributes to the useful impact resistant properties of the polyblend.

The second use takes advantage of the chemical composition

of the polymers [17]. Two polymers made from monomer A and B
might have relaxation peaks at quite different temperatures,
i.e. quite different values for the T_β peaks, Figure 3-3.
Copolymers of A and B could have intermediate T_β values. By
making a series of copolymers and mixing them one could get a
mixture with T_β drifting over a wide range of temperatures, or
test frequencies. Thus these polymers could be used to absorb
energy over wise ranges of frequency and to act as noise abatement materials. The specific examples which have been made
and do perform so are chemically heterogeneous copolymers of
vinyl chloride and 2-ethylhexyl acrylate in which the ratio of
the constituents could vary from pure vinyl chloride to pure
2-ethyl hexyl acrylate in individual polymer molecules. Polymers of this type can be made either using monomers with
suitable reactivity ratios or by incremental feed of one of
the monomers to the reaction mixture.

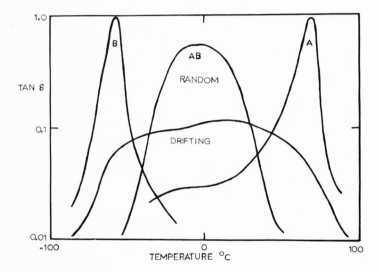

Fig. 3-3. Schematic representation of dynamic mechanical loss
plots for a hard polymer, A; a soft polymer, B; a random copolymer, AB; and a copolymer of A and B whose composition is
allowed to drift. As the peaks broaden they normally also
become lower but this may be experimental technique.

This does not exhaust what could be said about the glass transition temperature and the other relaxations which take place in the glassy or amorphous melt state. For much more detailed discussions, the various reviews of Boyer [17,18,19,20] and the monographs by Haward [21], Williams [8] and Schultz [23] are recommended. The importance of T_g in some other processes of engineering importance will be mentioned later.

4

The Crystalline State

Previously, mention was made of incipient order in amorphous polymers and melts. When the polymer molecules are regular in structure and there are regularly spaced and stronger attractive forces such as H-bonding, dipole-dipole, or favorable van der Waals forces the polymers will form crystals, that is, undergo crystallization [5,7,8,23]. There is nothing particularly different from the crystallization of small molecules, if you allow for the necessity of accommodating the long chain molecules. How this is accomplished is interesting.

Historically there have been two main theories of the structure of polymer crystals, or crystallites since they are often quite small and clumped into greater structures. The fringed micelle or oriented molecule theory is the older and applies to polymers in the solid state and to drawn fibers. In it a mole-

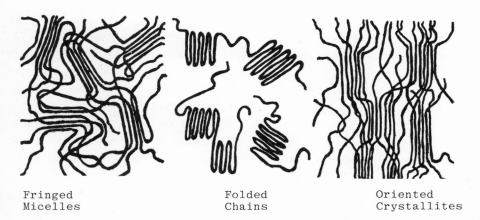

Fringed
Micelles

Folded
Chains

Oriented
Crystallites

cule will pass through several small crystallites formed where a number of chain segments become sufficiently organized to yield a crystallite. Each crystallite is like a bundle of twigs held in the hand, they are parallel within the grasp but

diverge in all directions at each end. The second theory is
the folded chain theory in which the polymer chain folds back
and forth on itself. Although such structures exist in solid
polymers they are best observed when polymers are crystallized
from solution to yield single crystals. Just exactly what
happens when the molecules turn is not settled yet but there
are several possible arrangements and perhaps several exist
simultaneously. The surface free energy of the side of the
crystallite is lower than for the end which would be related
to the turning-over of the chains, but the length of the fold

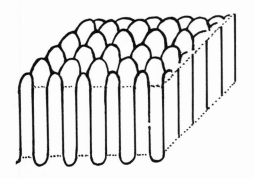

Schematic
Representation
of a Section
of a Folded
Chain Lamella

is not yet explained satisfactorily. The length does vary
with temperature, for example, but is surprisingly constant in
the range of 100-200 A. No doubt in systems as complex as
polymer molecules there are other ordered structures existing
which are crystalline in nature. The surprising thing is that
crystallization can take place at all. One must remember that
there is both crystallization, the actual formation of crystals,
and crystallizability, the ability to form crystals under
favorable conditions. The fraction of the polymer in the
latter category may well exceed that in the former by a considerable amount. Since polymers are seldom 100% crystalline
the term semi-crystalline polymers is usually applied. The
dividing line between amorphous polymers and semi-crystalline
polymers obviously must be vague since polymers which crystallize little or very slowly may appear amorphous normally.

Nucleation of Crystallites

The initiation of crystallization, or nucleation, takes place in two ways. The first is heterogeneous by which some inperfection or impurity acts as a locus about which the first orientation of the long chains takes place. These impurities may be chain ends, additives, impurities of many types, or deliberately added nucleating agents. The number and location of the crystallites will depend upon the amount and distribution of these nucleating sites. Purification of the polymer will reduce the number of sites. The other cause of nucleation is called homogeneous because it occurs uniformly throughout the mass of the polymer. There is usually an induction period during which the chains orient themselves into the beginning of a crystallite and then crystallization proceeds. Shearing the polymer or stressing it will tend to align the molecules and thereby induce crystallization, a fact to which we shall return later.

Growth of Crystallites

The growth of the crystallites is by a process of secondary nucleation which means simply that more chains add onto the nucleus gradually building the small crystallites into either a fringed micelle or a folded chain type of structure. There will be uncrystallized amorphous material in between which may be crystallizable but cannot move into the proper position or may be truly amorphous. In either case it will behave like the amorphous polymers described earlier, turning to a glass at its glassy transition temperature.

Crystallization can be followed in various ways but classically it has been by dilatometry in which the volume change on crystallization is measured in a dilatometer and the crystallinity calculated from the known densities of the crystalline and amorphous domains at the experimental temperature. A great many studies have been made and will be found in detail in monographs on polymer crystals and crystallization, particularly that by Geil [24] Only two observations will be quoted here. The first is that the rate of any crystallization tends

48

to be slower as the molecular weight increases as one would
expect since the first crystalline regions would restrict the
rearrangement of the remaining parts of the molecule. The
other concept is that the rate of crystallization is a maximum
between the melting point where of course it is stopped and
the glass transition temperature where on the other hand the
viscosity becomes too great for the molecules to align within
a reasonable time. A diagram illustrating this point based on
the data for natural rubber is in Figure 4-1.

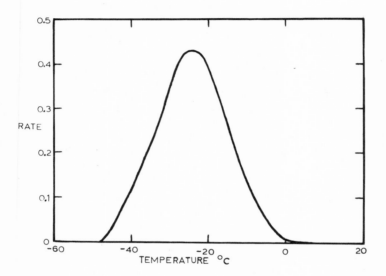

Fig. 4-1. The rate of crystallization of rubber plotted as the
reciprocal of the time required for one-half the total volume
change [27]. T_g would be about -70°C and T_m approximately 28°C.

As for simple molecules, the X-ray technique is the one
most commonly used to measure the dimensions of the unit cell
and the repeat distances along the chain. It is sufficient to
record here that polymers in crystallites exist in helical
conformations, planar zig-zag, etc. Only a very small part of

each molecule is in a given unit cell and several molecules
may pass through a unit cell. The determination of the crystal
structure of a new material is one well worth leaving for a
specialist. The same technique is used to study crystal-
crystal transformations both above and below the true melting
point, the latter being the $T > T_g$ transformation which often
has the properties of a crystal transformation. The size and
shape of crystals in a transparent amorphous or glassy matrix
may be measured by a light scattering technique which is based
on the scattering of a beam of light incident on the sample,
the intensity and the direction of the scattered light being
governed by the size and shape of the crystallites and the
difference in refractive index between the crystallite and the
amorphous phase. The fact that one may obtain a distribution
of crystallite sizes and perhaps shapes makes the technique not
an easy one. Of practical importance is the fact that the
light scattered by the crystallites leads to the phenomenon
described as opacity which may be desirable or undesirable.
It may be eliminated by making the crystallites small enough
by rapid quenching of the polymer so that any crystallites
which are formed are too small by the time the polymer is
cooled to the glass transition temperature to scatter visible
light.

Spherulites

The crystallites are usually small and often thin-plate or
rod-like. However, they are organized into various larger
structures which are easier to see and study. The single
crystal, folded-chain type crystals formed from solution build
into shallow pyramids composed of layers or lamellae. The
pyramids may be quite large and composed of a great many
layers or lamellae. The fringed micelle crystals formed from
the melt will differ in structure depending upon the crystallization conditions but if crystallization is allowed to proceed
normally under good experimental conditions a large structure
called a spherulite is formed. The name is derived from the
fact that it is an approximately spherical arrangement of
crystallites.

The morphology or structure of the spherulites has been studied extensively [24-26]. One reason is that it is relatively easy by using polarized light to observe the orientation of the anisotropic crystallites unlike the isotropic amorphous polymer. Thus one observes patterns with polarized light and crossed nicols and these patterns can in turn be related to the structure of the spherulite.

For example the normal, small spherulite gives an optical pattern which for an isolated one resembles a Maltese cross.

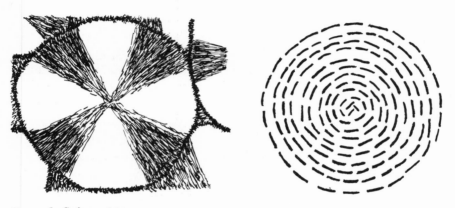

Normal Spherulite

This can be explained if the crystallites are arranged tangentially around the nucleus like the layers of an onion, although the structure is actually more like ribbons of crystallites, the ribbons repeatedly branching to yield the final spherically symmetrical structure.

A diagram of such a spherulite as it might be seen using polarized light and crossed nicols and how the crystallites might be arranged will illustrate the idea.

Less common very large spherulites have a root-like or vein-like appearance using polarized light and crossed nicols. Because of the similarity to roots and veins the structure is called dendritic. The crystallites are arranged in ribbons with many branches more nearly parallel to the radii and give rise to a herringbone-like structure. These concepts can be illustrated by simple diagrams also.

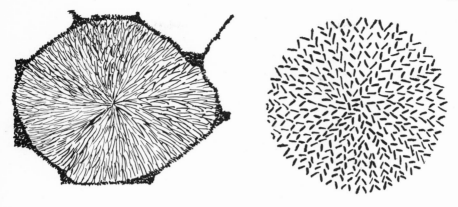

Dendritic Spherulite

Then there is a combination of the normal and "dendritic" types which appear to be a Maltese cross superimposed on a ringed structure when polarized light and crossed nicols are used, i.e. layers of crystallites alternately tangential to and along the radii of the spherulite. The cause for the twisting of the direction is not known exactly. This can be illustrated by simple diagrams but original photographs are needed to appreciate fully the arrangements in ringed micelles.

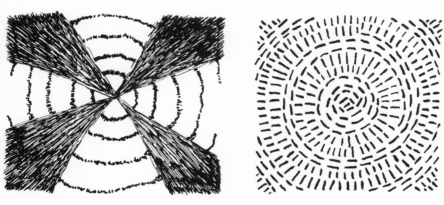

Ringed Spherulite

Finally, we shall mention hedrites or twin crystals, crystals with one common face. These occur for small molecules so that it is not surprising that they may occur for polymers also.

Since spherulites are composed of oriented crystallites
those factors which affect crystallite formation and growth
also effect spherulite formation and growth.   There are a
number of unusual features however.   Firstly the rate of growth
of the radius is linear, that is, the volume of the spherulite
increases with the cube of time.   Likewise the number of
spherulites increases linearly with time.   Of course the
increase in number and growth rate cease when the polymer has
reached its final or equilibrium state of crystallinity.
Growth slows and stops quite rapidly when this condition is
reached.   The spherulites when packed together take on various
shapes with flattened sides and perfect spherulites are only
seen in slightly crystallized samples.

Spherulites may form during polymerization particularly of
such monomers as ethylene and propylene which yield regular
structures with sterospecific catalysts.   Cellulose also is in
the form of folded chain crystallites as produced in nature
and the fibrids would correspond to spherulites extended in
one direction.   Just as for the amorphous polymer the structure of spherulites and crystallites appears to remain in
spirit during melting and processing of polymers and may
reappear very nearly as they were before the melting.   This
was quoted earlier as evidence for a tendency towards structure even in the amorphous and molten states.

The size and number of spherulites is of importance.   If
the spherulites are small and numerous the polymer tends to
be stronger, stiffer, more transparent, and more resistant to
failure.   If the spherulites are large there tends to be built-
in strains at the interfaces with the amorphous polymer which
results in crazing or silvering, the large crystallites may
fracture along the weak layers between crystallites and around
spherulites, and the polymer is more opaque.   The crystallinity
and hence the number of spherulites may be reduced by rapidly
cooling or quenching the polymer to below the glass transition
temperature and both the number and size can be increased by
annealing at a temperature between the melting point and the
glass transition temperature, perhaps at a temperature at
which crystallization is reasonably rapid but at which nuclea-

tion of new centres is not too rapid.

Melting of Crystals

Let us now turn to the reverse process of melting of
crystals. These melt just like any other crystalline solids
but entropy factors are very much more dominant, i.e. the conversion of the crystal to a random structure may be so inhibited by the stiffness of the chain and the viscosity of the
melt that true melting takes place at surprisingly high temperatures and the crystalline form and strength may persist to
temperatures well above that at which crystallinity as measured
by volume change and by X-rays has disappeared. Data to illustrate these factors are not easy to find but the following [5,7]
will illustrate the main points.

Table 4-1

Heat and entropy of fusion of polymer crystals

Polymer	ΔH_m cal/g.	ΔS_m cal/g.K	$(\Delta S_m)_n$ cal/g.K	T_m °C	T_g °C
Polyethylene	66.2	0.162	0.126	138	-120
Polyoxymethylene	59.6	0.131	0.079	180	
Polytetrafluoro-ethylene	13.7	0.023	0.015	327	130
Cis-Polyisoprene	15.3	0.051	0.028	20	- 70
Trans-Polyisoprene	45.3	0.130	0.075	74	
Polypropylene	10.7			165	
Polypropylene	35.0			183	26
Poly(4-methyl-pentene — 1)	33.9			250	130

The high melting points are due primarily to the low entropy
of fusion resulting from the stiffness of the chains. As the
structure approaches polyethylene so do the enthalpies ΔH and
entropies ΔS of melting.

The importance of the change in entropy of the melting of

polymer crystals is admirably illustrated by the effect of
thermal history on the melting range of natural rubber [27].
Natural rubber may be crystallized over a fairly wide temperature range, -40°C to 15°C. When samples are then melted,

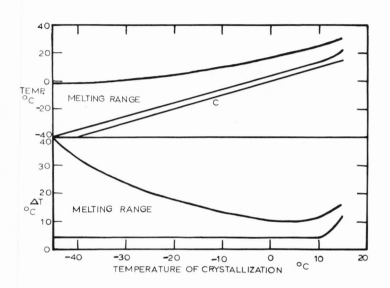

Fig. 4-2. Melting range of crystalline natural rubber, upper,
and the interval between the beginning of melting, and the
completion of melting, lower, as a function of the temperature
of crystallization. Line (c) designates the temperature of
crystallization relative to the temperature of the beginning
and completion of melting.

melting starts about 5° above the crystallization temperature
and continues over a range which is widest, nearly 35°, when
crystallization takes place at the lowest temperature   yielding
a range of imperfect crystals and decreases to a few degrees
when the crystallization is accomplished at the highest temperature and the crystals are more uniform. The data suggest
an equilibrium melting point of about 28°C which agrees with
the fact that "racked" natural rubber, which has crystallized

slowly at moderately low temperatures, remains crystalline at room temperature.

Drawing of Semi-crystalline Polymers

The crystalline regions in semicrystalline polymers may be disrupted by a process of cold drawing or distortion which orients the crystallites in the directions of drawing or elongation [5,7,23,28]. At the same time the chains in the amorphous regions are also oriented in the directions of drawing. For a fiber this is usually in just one direction of elongation, along the length of the fiber for strength. For a film or sheet the elongation may be made in two directions simultaneously resulting in biaxial orientation of the films. These films are useful for shrink-wrapping and as a strong film for packaging applications generally. Description of the process in one direction or dimension is adequate to understand it. Strength is imparted in the directions of draw but the strength in other directions will be decreased. If a simple diagram of crystallites is made and then an imaginary force applied to it the crystallites will become distorted, yield, and disrupt as shown progressively in the figures below.

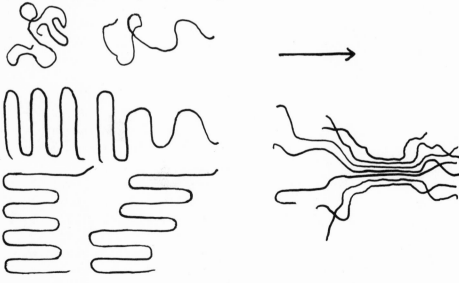

Orientation by drawing

Finally the polymer chains will become oriented in either a
fringed micelle or folded chain type crystallite or will remain
as undisturbed crystallites which happen to be lying with their
strong dimension in the direction of draw. The final result
is a stronger fiber or film.

Characteristic necking is observed when a fiber is drawn.
The large fiber narrows down relatively quickly with the evolution of heat, then the smaller fiber remains comparatively
unaffected in size under constant conditions. One can
visualize the changes taking place. In the smaller stronger
fiber the molecules and crystallites are oriented so that the
direction of greatest strength is along the fiber. This strong
fiber pulls material out of the larger fiber at the neck with
the evolution of the heat of melting. This heat is reabsorbed
on crystallization and orientation in the smaller fiber. A
diagram illustrating the necking of a fiber follows.

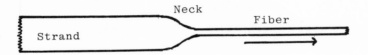

The same general changes take place in the spherulites. They
are at first intact, then become elliptical and finally melt
or disrupt and recrystallize into a stable form. The distortion and reforming of both crystallites and spherulites takes
place rapidly. The structures are somewhat different when the
sample is cold-drawn as above or hot-melt extruded [22]. Whereas the polymer molecules are aligned in the direction of the
stress in cold drawing, when the melt is extruded they are
at right angles to the direction of flow. An extruded mono-
filament must be subsequently drawn to have longitudinal
strength. Likewise oriented film can be made by drawing the
extruded film. Biaxially oriented film which is made in large
quantities is first extruded then drawn in two directions by a
variety of processes the best known of which is blown film
extrusion. Several other processes are used which draw the
film in two directions.

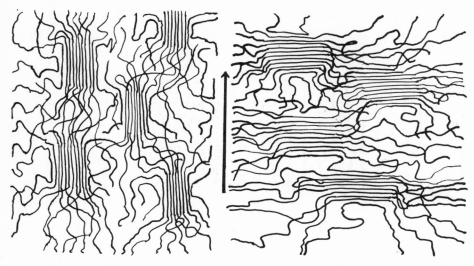

Cold drawn film   Extruded film

Tie Molecules

The preceding discussion implied that the lamellae making
up crystals and spherulites were independent like the layers
of an onion. However, in fact some molecules pass from
lamella to lamella in a single crystal and from crystallite
to crystallite in spherulites. These tie molecules hold the
structures together. They have become the source of considerable study. Obviously the number of tie molecules will affect
the ease with which the crystallites separate under stress.
Unfortunately there does not appear to be a good means of con-
trolling the number and location of tie molecules at present
except by the indirect process of drawing on the one hand and
the random cross linking by radiation which will result in the
joining of molecules in adjacent lamellae or crystallites.

There are a number of practical engineering aspects to
crystallization which should be mentioned at least briefly.
For example a filler in the amorphous phase may initiate
crystallization and thereby stiffen, harden or strengthen the
polymer just like a reinforcing filler. The properties may be
proportional to the induced crystallinity. On the other hand

the filler may restrict drawing and orientation and weaken a
mixture leading to sudden brittle failure under tensile stress.

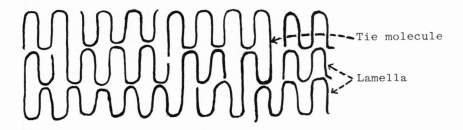

As mentioned earlier higher crystallinity with larger spherulites and crystallites yields a more brittle product than
smaller spherulites and crystallites and lower crystallinity
unless the orientation is favorable and ample numbers of tie
molecules are present. The forces within crystals are large
but between lamellae and the layers of spherulites the forces
are weaker. It is well to mention again that depending upon
the uses one may try to obtain many small crystallites and
spherulites or a few large crystals and spherulites, the latter
resulting in weaker products with more built-in strains at the
interfaces between the crystalline and amorphous phases.

Some Uses of Control of Crystallization

An understanding of the formation of crystals and spherulites and of the consequences of the unique properties they
possess is of great value in some practical problems. Only
a few may be noted. When a pipe is extruded, cooling the
inside results in crystallization of the inner layers first.
Then as the outer layers cool they shrink onto the inner
layers in effect yielding a stress-strengthened product in
which the outer layers exert a pressure on the inner layers.
This built-in extra strength against bursting may be lost due
to creep in due course but nevertheless is much better than
the reverse process of cooling the outside and allowing the
slower cooling inside layers to shrink away from the outer
ones leading to delamination and weakness.

Certain monomers such as caprolactam used to produce Nylon 6 can be polymerized in a mould and by judicious choice of conditions the crystallites can be oriented preferentially to yield stronger products. Preformed polymer used in rotational molding and other forms of extrusion, drawing, moulding, etc., can be oriented so that the strength of the product is in the desired direction. In the case of rotationally moulded articles this will be in the plane of the surface in all directions. However, orientation cannot be controlled always. Polypropylene crystallized in contact with chrome plate, a common mould surface, forms a columnar structure perpendicular to the mould surface and this can lead to the formation of surface cracks which, as will be discussed later, can cause mechanical failure by stress cracking.

Side Chain Crystallization

Sometimes the uniformity of structure and the cohesive forces favoring crystallization are present in the substituents or side groups rather than in the backbone or main chain. Under such conditions the side chains will form crystallites with the main chains remaining amorphous as for example:

Side chain
crystallization

Relation between T_g and T_m

At one time it was believed that T_g was half T_m on the absolute scale for symmetrically substituted chains such as polyisobutylene or poly(vinylidine chloride) and 2/3rd T_m for asymmetrically substituted chains such as polypropylene and poly(vinyl chloride). However, this rule is only an approximation and a guide for locating the expected values. Accurate data to compare homologous series of polymers are not readily available but the tendency can be noted in the data available for related polymers. However some polymers which do not seem to crystallize at all really have very high ratios of T_g to T_m. For example the ratio is reported to be nearly unity for poly(phenylene oxide) so that very exact control of conditions and times is required to form crystallites of this polymer, conditions not likely to be met in normal commercial practice.

5

Adhesion and Autohesion

One of the first engineering properties which illustrates
well the interplay between structure and morphology without
involving much prior treatment either from a theoretical or
technological point of view is adhesion [29,30]. At the same
time the reverse of adhesion, lubrication considered as the
opposite of friction [31], when the surfaces are moving, may be
included. When the two surfaces are the same adhesion may be
called tack or cohesion, frequently called autohesion to
differentiate it from adhesion when two different materials are
used.

Autohesion

Tack or autohesion, i.e. the adhesion of one surface to
another of the same composition, is the easier to visualize.
If two blocks of a polymer are pressed together there is at
first a dividing surface which eventually and ideally disappears to yield a single block.

  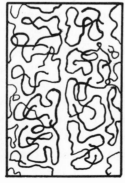 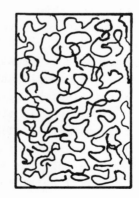

Obviously the strength of the adhesive bond ultimately should reach the cohesive strength of the material itself; the strength of the bond and of the bulk material should be the same. Many factors affect such a process whereby the tear strength of the interface is equal to that of the bulk, the cohesive forces as expressed by the cohesive energy density are maximal.

The process can only be complete if there is self-diffusion, a true diffusion of chain segments across the interface to obliterate it. Thus as the viscosity of the blocks of material increases so does the difficulty of flow and hence the auto-hesion decreases with increasing viscosity and increasing molecular weight and increases with the presence of solvents, plasticizers or tackifiers. Likewise the strength of the joint will increase with time, pressure and temperature of contact, factors which will facilitate knitting of the interface. The thickness of the samples or blocks has little effect except below a small value at which the nature of the substrate and its flexibility will interfere and will result in anomalous data.

In general the process of adhesion or self adhesion can be viewed as one involving self-wetting of the two surfaces followed by interdiffusion of the surface layers or as mutual dilution of the two surfaces. Thus in a thermodynamic sense one can treat the problem as one of heats of wetting and/or dilution and visualize the molecules which are normally con-strained by being at the surface having an additional dimension for attaining a more random conformation by diffusion across the interface into the other surface. Thus in addition to the factors mentioned above there are others which become important. Obviously the shape of the molecule will have an effect. If it is branched or crosslinked the diffusion process will be inhibited or stopped entirely. If the polymer is glassy or semicrystalline the diffusion will be inhibited or stopped but on the other hand if the cohesive join is made in the melt and then the polymer is allowed to form a glass or to crystallize then the resultant join will be strong. If there are surface layers which hold the samples apart the formation of the cohe-sive join will be slowed or stopped.

Tests for Autohesion

Time need not be spent on the methods of testing the
strength of cohesive or autohesive joins but these fall into
two main categories. The first is to press two samples
together for the desired time, at the desired pressure and
temperature, and under any other superimposed experimental
conditions such as surface treatments, etc., and then apply a
force to pull the two pieces apart again. This tensile pull
expressed in appropriate units, say kilograms per square centimeter of surface in contact can then be used as a measure of
the strength of the cohesive bond. A more common test is to
similarly put two strips in contact and then pull them apart
or peel one from the other. As is usual with most test methods
the experimental conditions must be standardized. For example
the measured peel strength or cohesive strength increases with
the rate of separation of the clamps applying the force, i.e.
with the rate of testing. In addition all the factors above
are dependent on time, temperature and pressure of the initial
contact of the two surfaces. Uniformity of treatment is
necessary to avoid discrepancies due to differences in the
morphology of the bulk polymer and cleanliness of the surfaces.

Factors Affecting Autohesion

The importance of tack or self adhesion and the attainment
of a strong cohesive bond between two samples of the same
material is not appreciated or recognized as much as the
adhesive process involving a second surface even though in this
case cohesive bonding may be an important factor. However,
before describing briefly a few important processes depending
on cohesive joins the factors which contribute to good joins
may be listed.

Since in general high molecular weight polymers have high
cohesive strengths and better properties but low tack, the
degradation of some of the molecules to yield a low viscosity
fraction which will diffuse across the interface may be of
assistance in forming a cohesive bond. This is practiced in
the mastication of rubber for which a high degree of tack is

required to build composite products such as tires which must be held together by tack during forming until the rubber is cured to give the final cured product with cured adhesion between the components. The tack may be improved by the inclusion of low molecular weight polymers or other materials such as solvents, plasticizers or tackifiers which fulfil the same function but of course may well reduce the final cohesive strength. Likewise the use of inert fillers such as kaolin, clay or silica or reinforcing fillers such as the very fine silicas and carbon black may have effects which are difficult to predict. In general if the filler is truly inert the autohesion will be reduced since there is less polymer surface in contact and the viscosity of the mixture is increased. On the other hand a reinforcing filler may, because of its fine nature and cross linking properties, result in better over-all contact of the two surfaces and form bonds across the interface yielding a better join.

The common practices in fabrication of articles in which two surfaces of the same material are to be joined include freshening or cleaning the surfaces, and roughening the surfaces to yield hills which come into contact in spite of an otherwise poorly fitting contact area. While an increase in the time of contact and the pressure is obvious, comparatively little can be done to change these. The temperature or its equivalent, i.e. fusion sealing or heat sealing, may be increased by application of dielectric, ultrasonic, thermal impulse or radio frequency energy which momentarily brings about considerable self-diffusion through localized heating, at the interface particularly if some polar material such as very finely divided iron oxide powder responsive to the energy input is used at the interface. These latter processes may be used when there would be loss of properties with loss of crystallinity or orientation on fusing the material by normal conductive heating. The thin film affected does not weaken the bulk of the sample appreciably. The cohesive strength increases with the polarity of the polymer when it bears groups such as CN, COOH, OH, etc. The same effect can be achieved by treating the surfaces with high energy discharges including electron beams, corona discharges, or just a flame to form an

oxidized surface with many polar groups present. If a solvent
is undesirable in the first product it may be possible to
soften just the surface and after the cohesive join has been
made allow the solvent to evaporate. Finally as mentioned
above the curing of the pieces in contact whether by chemical
means as in the vulcanization of rubber with sulfur or by the
cross linking of polyethylene by gamma rays could be used.
These processes form covalent bonds across the interface with
about the same frequency as within the bulk of the material
leading to a so-called cured adhesion which approximates the
cured cohesive strength of the material. Curing can be considered a means of raising the viscosity of the material to
a very high value i.e. thermosetting, after shaping or forming
under thermoplastic conditions.

Although friction will be mentioned later as a property,
the well-known heating effect of friction can be used in
joining. By spinning the two parts of a plastic object in
contact until a thin layer on each part softens or melts and
then pressing them together, a rapidly formed cohesive seal is
formed without loss of properties of the bulk of the material
which has remained cool.

Still another technique in common use which involves the
adhesion and cohesion of the same materials is the welding of
plastics during which pieces of plastic, ducts or pipes for
example, are joined by molten polymer which fuses with a molten
layer of the pipe or duct during the welding process. This
process has been used with poly(vinyl chloride) frequently but
is more difficult when the polymer loses some desirable
property on melting and cooling.

Coalescence

One aspect of autohesion or tack of special importance is
the coalescence of particles in emulsions or suspensions of
polymers into films. In this instance the surface tension or
surface energy of the polymer plays a very important role.
Thus at the point of contact of two particles the force
tending to cause the particles to assume a dumbell shape may
be very great.

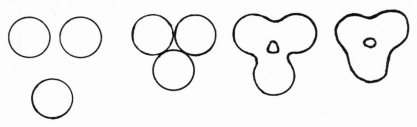

Flow to yield a smooth film may be very slow in the final
stages if the polymer is glassy or semicrystalline. It is not
appropriate to discuss the details here but it must suffice to
say that when the surface energy is great enough to overcome
the viscous forces within a single particle, flow will result
and the particles will fuse. Various factors affect this rate
of fusion. The amount and type of surface stabilizer will
affect the ease with which the individual particles come into
contact and destruction of the stabilizer may be necessary.
Time of contact, pressure, shear forces, and heating above the
glassy transition temperature or the melting point may be
desirable.

If a monolayer of particles is laid down these tend to pack
in a regular hexagonal pattern if the particles are of uniform
size but otherwise pack together as closely as possible. At
least as an approximation this same process takes place in
three dimensions. Thereafter the fusion of the particles must
be somewhat similar to the sintering of metals or glass which
can be observed on a larger scale and hence is better under-
stood. The final sintering or fusion involves the same factors
as autohesion, segmental motion during which segments diffuse
across the interfaces, entanglement of the molecules across the
interface, a disordering of the polymer chains on the surface
of the polymer particle where their randomness was limited by
the presence of the surface, all of which are aided by polarity
of the surfaces, softening of the polymer by heat, solvent or
plasticizer, the application of pressure, and time to allow the
slow diffusion processes to take place.

It is appropriate to mention that not only can one fuse polymer particles but in the presence of pigments, fillers, etc.,

these make paints. There should be adhesion to the pigments
and fillers and we shall consider this a process of adhesion
to be discussed later, including mention of surface treatments
which may be required. Also dispersions or lattices of two
different polymers may be used and these will fuse together.
Since the individual particles tend to retain their identity
there will be interpenetrating domains of the two types of
polymers. This will also be considered a problem in adhesion.

Wetting of Surfaces

The subject of adhesion and adhesives is a vast one. Just
enumerating the number of items in everyday use in which two
similar or dissimilar materials are held together by an adhesive would be an enormous task. It may be true that if surfaces placed in contact were perfectly smooth on a molecular
scale, clean, and pressed together for at least a short time
an adhesive bond would be formed. However, in general these
ideal conditions do not apply so that adhesives in the form of
emulsions, solutions, or molten polymers are used to join the
pieces. The question is whether we can make any general sense
out of such a diverse topic. The answer is yes, in general,
but the choice of the best adhesive for specific applications
may often be by test, guided by experience.

There are a number of ways of formulating or explaining
adhesion but that most commonly accepted is based on the
wetting of the surfaces to be adhered, and the work of wetting.
The work of wetting or adhesion is the work required to separate the two surfaces which are cohering, i.e.

$$W_{ad} = \sigma_{1,2} + \sigma_{3,2} - \sigma_{1,3} \quad (ergs/cm^2) \qquad (1)$$

where $\sigma_{1,2}$ and $\sigma_{3,2}$ are the surface tensions (dynes/cm) of
the two liquids (1 and 3) in contact with air (2) and $\sigma_{1,3}$ is
the surface tension of the interface between the two liquids.
When polymers are used one or both may not be liquid so that
it is more exact to use the equation:

$$W_{ad} = \sigma_{1,2} (1 - \cos \theta) \qquad (2)$$

in which θ is the angle of contact of the liquid droplet
with the solid substrate.

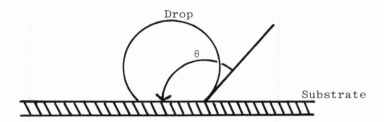

For complete wetting the angle is zero, for no wetting, 180°.
The above case is quite useful in studying the wetting of
solid surfaces by liquid adhesives but the most interesting
practical cases involve two solid phases. The work of adhesion for solids s_1 and s_2 in air (a) will be

$$W_{ad} = \sigma_{s_1 a} + \sigma_{s_2 a} - \sigma_{s_1 s_2} \tag{3}$$

The work of adhesion for a liquid and the two solids will be

$$W'_{ad} = \sigma_{s_2 l} + \sigma_{s_1 l} - \sigma_{s_1 s_2} \tag{4}$$

Subtracting

$$W_{ad} - W'_{ad} = \sigma_{s_1 a} - \sigma_{s_1 l} + \sigma_{s_2 a} - \sigma_{s_2 l} \tag{5}$$

By making separate measurements of the contact angle of a drop
of liquid on each of the solids and equating the two relevant
equations (1 and 2) yields at equilibrium two further equations

$$\sigma_{s_1 a} - \sigma_{s_1 l} = \sigma_{la} \; \mathrm{Cos} \; \theta_{s_1 la} \tag{6}$$

$$\sigma_{s_2 a} - \sigma_{s_2 l} = \sigma_{la} \; \mathrm{Cos} \; \theta_{s_2 la} \tag{7}$$

Substituting these in the general equation yields

$$W_{ad} - W'_{ad} = \sigma_{ls} \; (\mathrm{Cos} \; \theta_{s_1 la} + \mathrm{Cos} \; \theta_{s_2 la}) \tag{8}$$

If the surface free energy of the liquid, i.e. surface tension is reduced to such a value that the work of adhesion in the liquid phase is zero, which can be accomplished by using a series of solutions of surface active agents, the condition is reached for which the equation

$$W_{ad} = \sigma^0_{la} (\text{Cos } \theta_{s_1 la} + \text{Cos } \theta_{s_2 la}) \qquad (9)$$

holds in which σ^0_{la} is the surface tension of the liquid ensuring zero adhesion and equals σ_{ls}. This latter is conveniently obtained from a series of solutions with surface active agent present and the one in which two pieces of polymer no longer exhibit any pressure sensitive adhesion yields the limiting value for the surface tension. (see [207] in [29]).

Adhesion

Use of either a drop of liquid on a solid surface or the pressure sensitive contact adhesion of two pieces of the polymer immersed in aqueous solutions of surface active agent are each practical methods of estimating the surface free energy of the solids concerned and may lead to fruitful conclusions in understanding the adhesion of polymer to polymer and polymer to other substrates based on the wetting of the surfaces concerned.

Adhesion can also be considered an adsorption process although this is not a popular concept. The forces of H-bonding, active groups at the interface, chemical reaction between the phases and dipole and dispersion forces could all contribute to an adsorption-like mechanism. Ultimately the difference between adsorption and wetting, in practice, is difficult to resolve.

Still another view of adhesion is the electrostatic theory in which the surfaces are held together by electrostatic attraction. Credence was given to this theory by the fact that when two surfaces are pulled apart they are often found to be electrically charged. However, the charges are believed to result from the separation rather than to be the actual cause of the adhesion.

Closely allied to the theory of adhesive wetting is the viscosity concept. Two pieces of glass which are wet adhere very firmly but may be slid apart since the viscosity of the water is low. If the viscosity is increased greatly, this no longer can be done and so adhesion results. Thus if the adhesive wets the surface it may be applied as a liquid and solidifies to a high viscosity causing adhesion. These are so-called hot melt adhesives and are commonly used to seal cartons, etc. The higher the molecular weight or the viscosity of the cooled melt the stronger the adhesive, generally.

As with autohesion there are many factors which affect the adhesion or must be considered when the adhesive strength is considered. For example the coefficient of expansion of organic polymers is much higher than that for metals so that temperature changes tend to cause hard adhesive layers to be strained or fractured when they contract more than the metal on cooling. This in practice is overcome by using filled adhesives with lower coefficient of thermal expansion and making them sufficiently elastic that the strain imposed will not fracture them. In dealing with polymer-polymer adhesion the incompatibility of polymers must be recalled so that time must be allowed for the polymers to intermingle or assistance given to mixing the adhesive layer by using a gradation of properties or, as in the case of laminates, fuse the layers together while they are still molten by co-extrusion, extrusion coating or laminating.

Measurement of Adhesion

The measurement of adhesive strength can be made in various ways but the peel test mentioned earlier is the commonest and most useful. When the peel test is used various trends appear which parallel those obtained for autohesion. Time of contact, application of pressure, heating and then cooling, use of polar materials, use of solvents to freshen the surfaces (i.e. a common solvent) or to make the adhesive sticky and fluid to wet the surfaces, use of plasticizers or tackifiers, and use of low molecular weight polymer to wet the surface, etc. all

tend to improve adhesion. When all this is done it is often
found that the limiting strength of the adhesive bond between
substrate and adhesive, i.e. the adhesion to the substrate,
exceeds the cohesive strength of the liquid adhesive and there
is cohesive failure of the adhesive.

Since the substrate and adhesive are different, substrate
treatments may be needed to achieve good cohesion. There are
quite a few techniques applied to treat substrates. One simple
treatment is to clean the surfaces to remove air, water, oil
or other impurities. Another is to roughen it so there is a
mechanical interlock with adhesive. A third is to treat the
surface chemically. Metals may be phosphated or chromated.
Polymers may be treated with chlorine, ozone, isocyanates,
corona discharges, flames, or high energy radiation with the
purpose of putting active groups on the surface. These then
allow the adhesive to wet the polymer better and may ultimately
form chemical bonds with it. The well known treatment of glass
with silanes and silanols to bring about adhesion to polymers
in composites is such a case, the silanes remove water, become
attached to the surface and are compatible with the hydrocarbon
polymers which they wet upon contact.

Surface Coatings

A relatively new field of surface coating which is too large
and complex to discuss fully here depends upon adhesion and
autohesion. This includes electrocoating from aqueous solution, electrocoating from aqueous emulsion, electrostatic spray
coating, fluidized bed powder coating, and the many other
aqueous and non-aqueous systems whereby a finely divided powder
of a coating material is first adhered to the substrate or
surface and then, either spontaneously or by heating, the particles fuse together and adhere to co-sprayed pigments, fillers,
glass fiber, etc., by combined adhesive and autohesive
processes to give the many spray and powder coating processes
now in use.

Another new process of coating and hence of putting a layer
of adhesive on a surface, since a coated surface when pressed
together yields autohesion of the coating which then becomes

an adhesive, is the formation of polymers on a surface either
by direct electrocoating and polymerization of certain monomers,
or by initiating polymerization of a monomer on the surface of
one polymer, a process called grafting. In the former some
acrylic type monomers can be deposited on a surface from the
vapour phase and polymerized by electrical discharges and in
the latter the discharges will initiate the polymerization of
a monomer perhaps ethylene on the surface of polytetrafluoroethylene thereby yielding a surface much more easily wet by
inks or joined by heat and pressure than the pure polytetrafluoroethylene.

An interesting example of the hot melt type of adhesion is
a conical piece of polypropylene which is used to join polypropylene pipe ends formed into a cone and socket. The adhesive cone contains an electric heating element which upon
application of a current softens the polymer so that a good
join is made when two sections are pushed together. Then if
the pipe must be taken apart, the ends of the wires are connec-
ted to a power source again, the polymer softened and the pipes
pulled apart. Within limits the join is reusable and each time
gives a complete volume filling seal of polypropylene melt
which yields a strong and leak-tight seal on cooling with
strong adhesion and autohesion, due to the semicrystalline
nature of the polymers.

An important point to remember with adhesives is that
generally the strength increases with decreasing thickness of
the adhesive layer. Various theories have been suggested to
explain this. However, it is probably true that the adhe-
sively effective layer is very thin and once this is exceeded
the importance of the cohesion of the adhesive becomes increasingly dominant. As we shall discuss later the chances of
failure of such materials increases with sample size or thickness. In any case this is a good example in which some adhesive is good but a lot is not better and may indeed be worse.
There are, of course deviations from this, particularly when
sealing or void filling is part of good adhesion and then bulk
of polymer or adhesive may be necessary to reach the substrates
to be adhered. This process is closer to sealant or caulk
technology.

Friction and Wear

It is convenient to discuss friction [31] at this point since friction can be compared to a failure of adhesion or autohesion parallel to the surface in contact instead of normal. The science of tribology has become a very important one and only brief mention can be made of it. The friction of plastic materials is related to the viscoelasticity of the products and this will be discussed later. Briefly the friction of rubber on a solid surface is a stick-slip type of behaviour. The rubber sticks, then slips. This happens with a characteristic frequency which is related to the resilience or energy absorbing properties of the rubber. This in turn is a function of temperature as well as other variables, hence tires squeal on the roads at higher temperatures, and the more resilient rubber tires squeal more easily.

The other polymeric property of interest is the surface energy of plastics. For polytetrafluoroethylene the surface may be composed of rolled-up polymer chains which slide along with the object and lubricate its motion. Also the explanation for the low friction characteristics may be simply the low surface free energy so that matter in contact does not adhere [32], Table 5-1. In the case of nylon and other polymers which actively H-bond water to the surface, the layer of water acts as a friction reducing surface and one is dealing with the viscosity of water to some extent rather than the friction between nylon surfaces.

Finally one should note that composites of all types are being made whereby a low energy surface is exposed as the composite wears. There are polymer composites containing graphite or molybdenum disulfide, composites of polymers with silicones or polytetrafluoroethylene, foamed spongy or sintered products in which the voids are filled with lubricant, etc. Mention will be made of examples in many of the general references and certainly in texts on lubrication, and friction. In general we are concerned with reducing the friction. One should not lose sight of the fact that abrasive fillers or short wire may be added to increase friction, i.e. traction of tires for example.

Table 5-1

Effect of constitution on friction and wettability of halogenated polyethylenes

Polymer	Static coefficient of friction	Critical surface tension, dyn/cm
Poly(vinylidene chloride)	0.90	40
Poly(vinyl chloride)	0.50	39
Polyethylene	0.33	31
Poly(vinyl fluoride)	0.30	25
Poly(vinylidene fluoride)	0.30	25
Polytrifluoroethylene	0.30	22
Polytetrafluoroethylene	0.04	18

Abrasive Wear

When two surfaces do adhere even partially and are forcibly
rubbed together as in the sliding of a tire there is abrasive
wear during which chunks of the rubber are torn from the surface layers. This shows up as the black streaks on roads since
the outer wear-resistant layer of a tire is rich in the carbon-
black reinforcing agent. Thus adhesion, followed by cohesive
failure in the tire leads to abrasive wear.

6
Rheology, Viscous Flow, Elastic Liquids

Polymer melts and polymers in solution are frequently
studied under dynamic conditions which simulate in many ways
the behavior of the same polymers or solutions during fabrication processes or use. The intensiveness of the research
and its exactness perhaps may not compare with similar studies
applied to polymer characterization but the results and the
concepts have value in understanding the properties and perfor-
mance of materials, and as design data. Two broad aspects of
the subject will be considered, the thermoplastic state of
polymer melts or elastic liquids and in the next chapter, the
viscoelastic properties of elastic solids.

Viscosity and Viscous Flow

Viscosity or the resistance to flow of a melt or a solution
represents a frictional loss in energy which appears as heat
5,7,8,33-37. The definitions of the main variables can be
visualized in the diagrams. The left hand one represents
distortion by a shearing force (f) and the right hand diagram
represents flow of velocity (u) by a similar force (f).

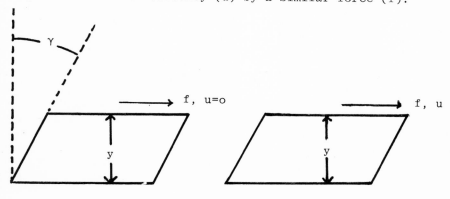

Static Shear   Dynamic Shear

The shear stress or force per unit area, say dynes per
square centimeter, is defined as τ. The shear rate (or rate
of shear) $\dot{\gamma}$, is the velocity of the movement of the plane where
the shear force is applied per unit of distance to the
stationary layer and has the dimensions of reciprocal seconds.
In the diagram it is du/dy. The shear strain (γ) is the
amount the plane where the shear stress is applied has moved
per unit of distance to the stationary layer. Being a ratio,
it is dimensionless. Finally the resistance to the flow or
shear viscosity (η) is the shear stress divided by the rate of
shear ($\tau/\dot{\gamma}$) and for most of the examples used here will be in
poises or centipoises. Two calculations can be made from the
above numbers. The first is the kinematic viscosity which is
the viscosity divided by the density, η/ρ. The other is the
rate of energy dissipation, $\dot{Q}$, which has various forms such as
τ^2/η, $\eta(\dot{\gamma})^2$, and $\tau\dot{\gamma}$, in each case ergs per cm^3 per sec.

To put the values of the shear rates in perspective the
shear rates for typical processing methods are for compression
molding 1 to 10 sec^{-1}, for calendering 10 to 10^2 sec^{-1}, for
extrusion 10^2 to 10^3 sec^{-1} and for injection molding 10^3 to
10^4 sec^{-1}. The above applies to polymer melts and the numbers
given are just typical values. The important idea to grasp is
that the shear rates are high as a result of the high viscosity
of polymer melts and the high energy input required to mix and
shape the products.

For the much lower viscosities of solutions of polymers the
forces are smaller and the technique is used frequently as a
means of characterizing polymers. Recalling polymer solutions,
a number can be measured which corresponds to the root mean
square end-to-end distance. This is proportional to the molecular weight of the polymer in solution and varies also with
the solvent. Under certain conditions the proportionality may
be ideal in the sense that the viscosity is proportional to
the square root of the molecular weight, an equation similar
to the Einstein equation. Such a solvent or solvent mixture,
or at such a temperature that a solvent will achieve these conditions, is referred to as the theta or Flory solvent and temperature respectively. In general the cohesive energy density
of the solvent for maximal swelling or maximal viscosity of a

polymer in solution will be such that the swelling will be
greater, and the root mean square end-to-end distance will be
greater, than for θ conditions. The ratio of the two values
is the expansion coefficient as described earlier.

Solution Viscosities

The viscosity of solutions and solvents may be expressed as
η and $η_s$ respectively, in poises. The ratio $η/η_s$ is the
relative viscosity η. Subtracting the viscosity of the solvent
yields the specific viscosity $η_{Sp} = η_r - 1$. Both of these are
dimensionless ratios. For more practical purposes the viscosity divided by the concentration is more often used, ln $η_r/c$
is the inherent viscosity and $η_{Sp}/c$ is the reduced viscosity
each with the units, deciliters per gram from the convention
of expressing concentrations in terms of grams per 100 milli-
liters. Finally a plot of either the inherent or the reduced
viscosity versus concentration extrapolated to zero concentration yields a viscosity at zero concentration which is defined
as the intrinsic viscosity, also in deciliters per gram. This
intrinsic viscosity, identified by the square brackets, is a
very useful measurement. Its importance lies in the relationship $[η] = K' M^α$ in which M is the molecular weight of the polymer and K' and α are constants which can be measured and tabu-
lated. The values vary but under theta conditions α is 0.5
hence sometimes it is desirable to make measurements with such
solvents or at such temperatures that the theta condition
applies. A table of representative values follows [5], Table 6-1.

Melt Viscosities

Since solutions represent the extreme case of a plasticizer
being present in quantities much greater than the polymer
there is interest in relating the two concepts; dilute solution
properties and plasticizer action. The effects of concentration (c) is illustrated by the equation log $η_{Sp}/c$ = log $[η]$
+ K" $[η]$ c. As (c) equals or exceeds 50, i.e. when the polymer equals or exceeds the solvent or plasticizer the viscosity
becomes independent of solvent, the plasticizer effect is

observed and the viscosity of the solution or plasticized
polymer equals the viscosity of the polymer times its volume
fraction.

Table 6-1

Constants for viscosity - molecular weight relationship

Polymer	Solvent	Temperature, °C	$K' \times 10^5$	α
Cellulose triacetate	Acetone	25	8.97	0.90
SBR rubber	Benzene	25	54	0.66
Natural rubber	Benzene	30	18.5	0.74
	n-Propyl ketone	14.5	119	0.50
Polyacrylamide	Water	30	68	0.66
Polyacrylonitrile	Dimethyl formamide	25	23.3	0.75
Poly(dimethyl siloxane)	Toluene	20	20.0	0.66
Polyethylene	Decalin	135	62	0.70
Polyisobutylene	Benzene	24	107	0.50
	Benzene	40	43	0.60
	Cyclohexane	30	27.6	0.69
Poly(methyl methacrylate)	Toluene	25	7.1	0.73
Polystyrene				
Atactic	Toluene	30	11.0	0.725
Isotactic	Toluene	30	10.6	0.725
Poly(vinyl acetate)	Benzene	30	22	0.65
	Ethyl-n-butyl ketone	29	92.9	0.50
Poly(vinyl chloride)	Tetrahydro-furan	20	3.63	0.92

The relationship between the viscosity and molecular weight
is more difficult to compare. We had the relationship between
the intrinsic viscosity, i.e. solution viscosity at zero concentration and molecular weight. At the other end of the
scale we can measure the melt viscosity versus the molecular
weight. A plot of the logarithm of the viscosity versus the
weight average number of chain atoms yields (Figure 6-1) two
straight lines which intersect at about log Z_w = 3 in most

cases. Z_w is the weight average molecular weight divided by the molecular weight per chain atom, or the average molecular weight per chain atom and hence is really a measure of the

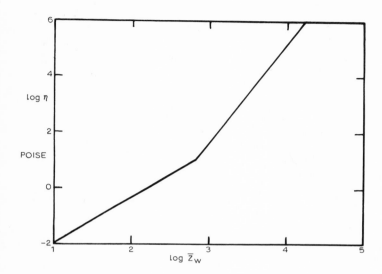

Fig. 6-1. Logarithmic plot of viscosity η vs. weight-average number of chain atoms Z (adapted from [5]). Break is interpreted as an entanglement length.

length of the chain. Equations can be written for the two portions of the curves and surprisingly the slopes of the lower portions lie between 1 and 2 and above the break between 3 and 4, in fact frequently very close to 3.4. The interpretation of this break point is that it represents an entanglement number. Above the chain length indicated by the break point the chains are entangled and have pseudo-crosslinks. This makes process-ing more difficult and indicates also when one might expect truly high polymer properties rather than highly viscous liquid properties, i.e. the appearance of elasticity, etc.

Since viscous flow can be viewed as movements of segments of molecules one could expect the viscosity to change with temperature and pressure similarly to changes in free volume. The changes with temperature can be formulated as $f' - f'_g =$

$\alpha\Delta T$ i.e. the fractional free volume, f', minus the fractional free volume at T_g, f'_g, is equal to the difference in the expansion coefficients above and below T_g (α) times the change in temperature. The value of (α) is about 4.8×10^{-4}/K. It is of interest that T_g will occur when f'_g is about 2.5% for polymers, many inorganic glasses and low molecular weight compounds. This is of course the equilibrium free volume and the actual measured free volume, and hence T_g, may be higher or lower depending upon the thermal history, etc. as described earlier.

In the more conventional terms $\eta = A\,e^{\varepsilon/RT}$ in which A is the frequency factor and ε is the energy of activation or temperature coefficient for τ or $\dot{\gamma}$. These equations hold over about a 100°C range normally. When ε_τ is measured this is at constant shear stress whereas $\varepsilon_{\dot{\gamma}}$ is at constant shear rate. The values of ε are about equal at low stresses but with increasing γ, $\varepsilon_{\dot{\gamma}}$ decreases and with increasing τ, ε_τ increases. Both would remain constant under Newtonian conditions.

As one would expect, high pressures decrease the free volume and increase the viscosity, hence T_g, for otherwise identical conditions. In general we assume the polymers to be incompressible and leave the details for research studies.

Non-Newtonian Flow

In the foregoing, the description implied that viscosity behaved in an ideal or Newtonian manner, i.e. the rate of flow or strain rate was proportional to the applied force. These assumptions are valid for most simple molecules and systems, and even for some polymer systems, but for most of the practical cases they are not valid. If the viscosity of a polymer melt or a solution is followed with time the viscosity for a Newtonian system should remain constant. If it increases it is called rheopexy, if it decreases it is called thixotropy.

If instead of time the shear rate is increased the Newtonian system should show direct proportionality between the applied stress and the shear rate since they are related by the equation $\tau = \eta\,\dot{\gamma}$. If the shear rate does not increase as rapidly as the applied stress the system is dilatant, if it increases more rapidly then it is pseudoplastic. This is the

usual form of the curve for plastics. Even so the curve may
have a shape which is Newtonian at low and high shear rates
and non-Newtonian at intermediate shear rates. If a finite
stress must be applied before any flow starts, i.e. a yield
stress, the system is called Bingham. After the yield stress
has been reached the material may behave Newtonian, dilatant
or pseudoplastic.

A much more complete discussion of rheology is better left
for a separate study. There are two more immediately practical
aspects of the subject to be included here. The first is an
outline of the methods of measurement, and the second is a
subject referred to as elastic liquids.

Measurement of Viscosity, Rheometers and Viscometers

The methods of measurement and instruments used are numerous.
They can be outlined under three ranges or types. The first
is the simple capillary viscometer such as the Ostwald, Cannon-
Fenske or Ubbelohde. In these, solutions of polymers flow
through standardized capillaries under gravity. The shear
rates are low but may be increased by applying pressure above
the surface of the liquid. One variation of this procedure is
to use a horizontal capillary between two reservoirs and to
drive the liquid back and forth until a suitable number of
measurements has been made.

This type of process can be used for concentrated solutions
and melts also either by using a ram to force the more viscous
contents through the capillary or by using a large orifice such
as the Ford cup test. The capillary type is more useful for
polymer melts in the form of the melt flow rheometers. For
less viscous materials various rotating cylinder types are used.
A cylinder or disc supported on a calibrated wire or spring is
rotated at various rates in the sample and the viscosity of the
sample is measured by the drag of the cylinder on disc on the
spring or wire. Viscometers of this type include the Brook-
field and the Haake. Alternatively a coaxial cup and bob may
be used (Contraves Rheomat-15) in which the cup or bob is
rotated and the drag on the other component measured by the
deformation of the standardized wire or spring.

When the viscosity of polymer melts is to be measured ram
extruders may be used. Two well known models are the Instron
Capillary and Monsanto Rheometers. The range may be extended
to quite high values of viscosity by the use of the cone and
plate or parallel plate rheometers such as the Weissenberg
Rheogoniometer, Mooney Plastometer, Rheometrics Mechanical
Spectrometer, Ferranti Shirley Rheometer, or Instron Rotary
Rheometer. These instruments not only apply quite high shear
rates to a viscous sample but also the Weissenberg or the
Rheometrics can be operated in an oscillatory mode for elastic
solids and are equipped so that both rotational and oscillatory motions may be applied simultaneously with relatively easy
changes in the rate of rotation and the frequency of oscillation. This by no means exhausts the list of types and uses
of rheometers. Many of the types mentioned can be used with
various types of sample holders. A practical instrument which
simulates commercial practice more closely is the Brabender
Plastograph.

Elastic Liquids

Recalling again that Newtonian conditions $\tau = \eta \ \dot{\gamma}$ implies
that the force is applied and the rate of shear is measured in
the same direction. If you used a cone and plate viscometer
you would note a pressure perpendicular to the surfaces of the
cone and plate tending to force them apart. Another anomalous
behavior is the stirring of viscous polymer solutions when it
may be observed that the contents of the vessel will climb up
the stirring rod. This is called the Weissenberg effect.
Still other manifestations of some different force are die
swell and melt fracture. These are illustrated by the series
in which as the pressure, and hence the shear rate, is increased
the polymer leaves the capillary first as a smooth rod the
size of the capillary, then swells on leaving the capillary,
then the swollen extrudate roughens to a so-called shark-skin
appearance which becomes more pronounced, irregular and rough
until a grossly misshapen "melt fracture" extrudate is formed
in which the rod appears to have swollen and fractured irregularly when the highest shear rate is used.

increasing flow rate

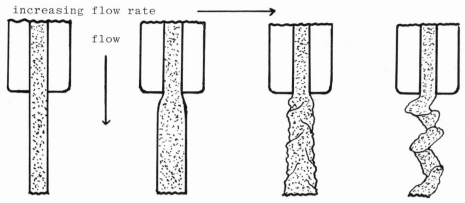

flow

Extensive studies of this phenomenon have led to mathematical analyses of which only the outline will be included followed by definitions. More details can be found in a monograph by Lodge [38].

If an element of a fluid is taken and the stresses on it analyzed they may be visualized as nine in number:

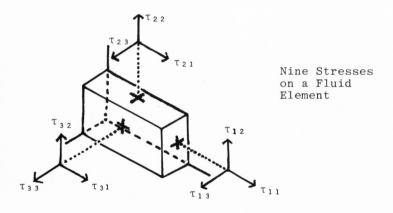

Nine Stresses
on a Fluid
Element

If one considers the pairs τ_{21} and τ_{12}, τ_{13} and τ_{31} and τ_{23} and τ_{32} these will represent opposing rotational forces which in a static system will be equal and opposite. They will be involved in shear forces, i.e. simple stress in the 11 direction will involve τ_{12} and hence τ_{21}. The terms τ_{11}, τ_{22} and τ_{33} in a static case represent hydrostatic pressures in the three directions and in a small enough volume of a liquid will

be equal. However, if one takes a shear stress τ_{12} it results
in a value $\eta\dot{\gamma}_{12}$, i.e. a rate of strain in the 12 direction.
This will in effect exert a pressure in the 12 direction and
this shows up as stress in that direction. This stress
$\tau_{22} - \tau_{11} = P_n(\dot{\gamma}_{12})^2$ is called the first normal stress
difference and this is the force that tends to push the cone
and plate apart and causes the viscous material to climb the
stirrer. It can be considered stored energy related to the
system by the product of shear stress and the recoverable shear
strain $\gamma_R = \dfrac{2\dot{\gamma}y}{x}$ i.e. $\tau\gamma_R = \dfrac{2\tau\dot{\gamma}y}{x}$. The recovery of the

stored energy accounts for the strain observed as the polymer
melt or solution rises, but more particularly as an extrudate
swells. Ultimately it causes the shark-skin, roughened and
melt fracture conditions of the extrudates.

There is a corresponding $\tau_{33} - \tau_{22} \cong 0$ second normal stress
difference but it is smaller under most conditions and requires
more research in order to be used to explain phenomena of
practical importance.

Drag Reduction

Another phenomenon which may be related to this concept is
drag reduction [39] whereby a small amount of a very high molecular weight polymer will reduce the viscosity of a liquid so
that it will flow faster. It is possible that the stored
energy is in the form of deformed polymer molecules and
surrounding liquid and these elastically reverse at the orifice
to exert additional energy in the direction of flow thereby
extending the range of a fire hose or assisting the flow of
oil or water through a pipe. Alternatively the energy absorption and release delays the onset of energy dissipating turbu-
lence during flow.

Applications

As mentioned earlier this type of viscous flow applies to
the processing conditions of extrusion, milling, molding, etc.

The viscosity appears to increase also with molecular weight
at constant temperature until entanglements form and then the
rate of increase of viscosity is greater. The viscosity tends
to decrease with temperature and increase with pressure in
keeping with the free volume concept and the changes in rates
of relaxation of segments of polymer chains with temperature
and rate of movement, i.e. shear rate. To explain some data
one must invoke still another concept which may well relate back
to the first normal stress difference. When certain polymers,
notably natural rubber [40,41] and poly(vinyl chloride) [42,43]
are processed there appears to be flow units which approximate
the original latex particles in size. Thus there appears to
be faults or regions around the original morphological structures which delineate flow units which persist through processing. Studies of other polymers may well show that they too
have a micro-turbulence which is frozen in and is responsible
for fracture and failure and when the polymer melt is sheared
are manifest by a normal force, a first normal stress difference
which holds discrete domains of the polymer together as flow
units.

Processing is related to the structure of the polymer and
probably to the molecular weight distribution. A study [44]
of polybutadienes with various end group using the Weissenberg
Rheogoniometer showed that polar end groups such as COOH or OH
increased the viscosity. The same study showed that this type
of approach could elucidate the effect of limited distributions
of molecular weight with different degrees of overlap of the
peaks on the viscous and elastic properties. This could be
related to the processibility of the melt.

Still further evolution of this rapidly expanding field of
rheometry is a study [45] in which the time - temperature superposition principle to be described in the next chapter was
applied to the flow of elastic liquids and the data used to
model viscosity for compression molding.

7

Viscoelasticity

In the preceding chapter we described the viscous flow of liquids and some aspects of the elasticity of liquids which were able to store and release energy. When the polymers become solid the elastic component may become dominant as in the case of cured elastomers. The viscous component is still present and particularly so in leathery or plastic polymers. There are three main types of stresses which may be applied to such viscoelastic systems [46-51] and it would be well to define the terms used [5,7,8,23,33] commonly to describe such properties of polymers. In this chapter we shall limit the discussion to small deformations not resulting in failure.

Firstly there is the system very similar to the elastic liquids previously described. A shear stress τ equal to the force per unit area is applied to a standard sized sample, say 1 cm cube. The stress would be expressed as dynes per square centimeter, for example. This results in a shear strain γ which would be the distance the surface upon which the shear is applied moves over the thickness of the sample to the surface of zero movement. Since this is a ratio it is dimensionless. Then the ratio of the shear stress to the shear strain is designated G (dynes per cm^2), the storage or shear modulus, for an ideally elastic sample. Emphasis on the words ideally elastic will be explained later. Sometimes the reciprocal of the shear modulus the shear compliance is used (dynes $^{-1}.cm^2$). The above corresponds to elastic liquids. Different symbols are used by deformations in tension so that later we can discuss viscous and elastic terms without confusion. Deformations in tension are often described in elementary courses. When a sample, say a rod, is stressed by some force (f) the tensile stress will be $\sigma = f/ab$ where (a) and (b) are the width and thickness of the rod, hence the product is the area of cross section. Again the stress will be dynes per cm^2, As a result of this stress the sample will elongate by an amount ΔL the ratio of which to the original length L_μ, i.e. $\Delta L/_{L\mu}$, is ε, the elongation. Since ΔL is the new length minus the original length it is often useful to convert ε to $\alpha -1$, i.e. the number of times the original length less one. Both ε and α are dimensionless and expressed usually as ratios to or multiples of the original length although the percentage increase in length is often used in technological literature, 300 percent elongation being an α of 3. The advantage of α is that the product of σ and α gives the stress per unit of actual cross section area of the elongated sample which numerically can be greatly different from the stress per unit of original cross section as is usually used.

As before the ratio of the tensile stress to the elongation is Young's modulus, $E = \sigma/\varepsilon$, dynes/cm^2, and the reciprocal is the tensile compliance J, ε/σ, dynes^{-1}.cm^2. When an elastic solid is stretched one would normally expect that the elongation would be compensated for by a decrease in the width and thickness. If this takes place precisely then Poisson's ratio is said to apply, the ratio designated ν is 0.5 ideally and mathematically is $-d(\ln a/d(\ln L)$ or $-d(\ln b)/d(\ln L)$.

The third system is the bulk stress in which a sample of volume Vo is subjected to an hydrostatic pressure P. The stress will be the increase in pressure, ΔP, (Pascals or kg/cm^2 or other units) and the strain will be the decrease in volume ΔV divided by the original volume, $\Delta V/Vo$, a dimensionless ratio. The bulk modulus (B) is then the ratio stress/strain, $\frac{\Delta P}{\Delta V/Vo}$ which is the reciprocal of the compressibility.

For isotropic materials, amorphous elastic or viscoelastic solids, $E = 2 G(1 + \nu) = 3B (1 - 2\nu)$. These are ideal conditions but by assuming Poisson's ratio of 0.5 at least a fair approximation of the other moduli may be obtained if the value of one is known.

Moduli are often measured on beams. If a beam is supported at one end only and a length L extends from the clamp, then a downward force (f) on the free end will deflect the beam downward a distance Y which is $Y = 4 f L^3/ba^3 E$ in which (a) is the depth of the sample and (b) the width. Likewise if

the beam is supported at both ends on fulcrums a distance L
apart $Y = f\ L^3/4ba^3E$. By rearranging the equations (E) may be
calculated. Similar types of equations have been developed
for samples in torsion but these are so dependent on the sample
shape and experimental conditions that one should consult the
proper references on mechanics. The torsional system will be
mentioned again later under the subject of logarithmic decrement analysis of transition temperatures. The technique is
finding increasing use in the torsional braid analysis procedure.
Both tests are standardized to simplify the calculations.

   For the remainder of this course, deformations in tension
and in shear will be discussed. Also, it will be assumed that
the polymeric mixtures are incompressible and that Poisson's
ratio holds. Deviations from these conditions are met in practice but are largely of research interest.

Models for Viscoelastic Systems

   When a stress is applied to an elastomer by stretching,
compressing, or bending, i.e. straining in some way, work is
done resulting in a deformation or strain. Some of this work
appears as heat and some as orientation, or elongation of the
molecules in the direction of the stress. This latter represents a deviation from the most random conformation and hence
the sample will try to revert to that most random or most
probable conformation when the stress is released. The behavior
is similar to elastic fluids discussed earlier except that the
return to the original conditions is more complete.

   One of the ways which has been used to help one visualize
what is happening is to use Maxwell or Voigt models of
viscoelastic systems. These are composed of a Hookean spring
(which obeys Hooke's law) and a Newtonian (viscous) liquid
which behaves like a shock absorber or damping device. If the
stress is defined as σ, the elongation as ε and the tension
or tensile modulus as E, then

Maxwell Model   Voigt Model

$$\sigma_1 = E\varepsilon_1 = \eta\frac{d\varepsilon_2}{dt} \qquad\qquad \sigma = E\sigma_1 + \eta\frac{d\varepsilon_2}{dt}$$

$$\sigma_2 = \frac{\eta d\varepsilon_2}{d}$$

$$\sigma = \sigma_1 = \sigma_2 \qquad\qquad\qquad \sigma = \sigma_1 + \sigma_2$$

$$\varepsilon = \varepsilon_1 + \varepsilon_2 \qquad\qquad\qquad \varepsilon = \varepsilon_1 = \varepsilon_2$$

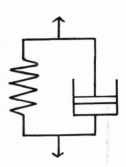

When the Maxwell model is used, the application of a stress will result in an instantaneous elongation of the spring, followed by a slow movement of the dashpot. The ratio of the viscous movement to the elastic movement for the overall process, i.e. when the stress and the elongation have reached their final equilibrium values, η/E, is defined as the retardation time.

A different model was proposed by Voigt. The spring and dashpot are in parallel. The horizontal lines at the top and bottom of the Voigt Model move in such a way as to always remain parallel. With this model, it is perhaps easier to visualize retardation time since the changes in E and η, which affect the ratio and the respective stresses born by elastic and viscous portions of the model, can be changed through extreme cases easily. For example, if E is large, and η small, the retardation time will be small; whereas, if E is small and η large, the retardation time will be large, in keeping with the relative rates of separation of the top and bottom lines under a given stress. For many discussions, it is convenient to use a Maxwell and a Voigt Model in series.

Of the many measurements made on viscoelastic systems, three will serve to illustrate static or once-through movements, elongation at constant rate of strain, creep, and stress relaxation.

The first mentioned is the most common method of measuring the stress strain properties of viscoelastic materials. The shape of the curve varies greatly, but a typical stress versus strain curve would be below with rupture or failure taking place at the end of the curve. The equation of the curve would be

$$\sigma = K_o \; \eta \left[1 - \exp\left(- \frac{E\varepsilon}{K_o \eta}\right) \right]$$

where K_o is the rate of strain, $d\varepsilon/dt$. Since for many viscoelastic systems the curves are more complex, a detailed study of all systems need not be made here.

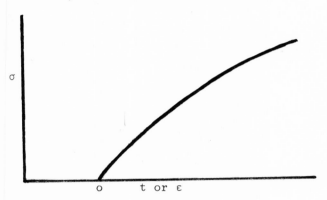

Creep is illustrated in the diagram in which a stress is applied at time o. The sample immediately elongates elastically then continues to elongate or creep until the stress is removed at time t. The sample immediately retracts elastically

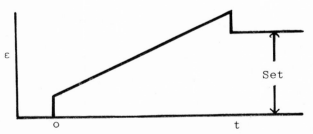

but remains elongated by an amount corresponding to the creep, the so-called permanent set. Since the equation for the portion of the curve representing creep is

$$\varepsilon = \varepsilon_o + \frac{\sigma_o t}{\eta}$$

the slope of this portion is σ_o/η. This illustrates clearly
that a high initial stress will increase the rate of creep,
whereas a high viscosity will decrease the rate.

Stress relaxation is somewhat similar. A typical plot of
ln σ versus time would be as follows.

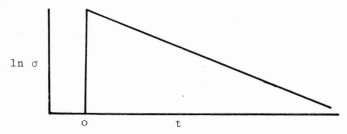

The equation of the curve would be represented by

$$\ln \frac{\sigma}{\sigma_o} = - \frac{Et}{\eta}$$

and the slope would be $-E/\eta$. Under these conditions, the
ratio η/E is the relaxation time, a measure of the rate at
which the stress decays at the constant strain. Recall that
earlier it was called the retardation time.

Dynamic Systems

Of particular interest is the response of a viscoelastic
material to a harmonic motion, dynamic mechanical stresses.
If a sinusoidal stress is applied to one end of a sample, the
response at the other end may be out of phase with the input,
the difference being measured by a phase angle δ.

Furthermore, Young's modulus for a system in tension can be
considered a complex function (E*), the ratio of the complex
stress σ* to complex strain (ε*) and the sum of an inphase
elastic or storage modulus (E') and an out-of-phase loss
modulus (E'').

92

$$E* = E' + iE'' = \frac{\sigma*}{\varepsilon*}$$

Mathematical analysis shows that the tan δ is E''/E'
(or J''/J'), the loss factor; i.e. the ratio of the loss to
storage modulus. Similar relationships exist for shear conditions.

$$G* = G' + iG'' = \frac{\tau*}{\gamma*} \text{ and } \tan \delta = \frac{G''}{G'}$$

E'' or G'' are measures of the energy lost per cycle and are
known more familiarly as hysteresis.

Master Curves, the Superposition Principle

Viscoelasticity may be divided into five regions, glassy,
glassy transition, rubbery, rubbery flow, and liquid flow. It
is not easy to measure viscoelastic properties over the wide
range of conditions to find all of these stages. Fortunately,

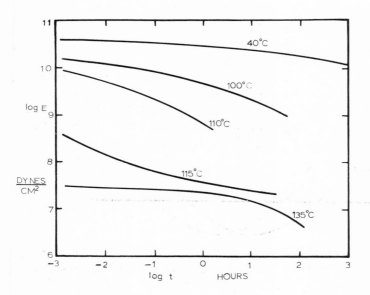

Fig. 7-1. Log E_r vs. log t for unfractionated poly(methyl
methacrylate) of $M_v = 3.6 \times 10^6$ after [5,52].

the Boltzman time-temperature superposition principle may be applied to obtain very useful master curves or generalized relationships.   As experimental work progresses further and further in both directions along the time axis, the value of the superposition as a useful tool has continued to be enhanced.

The data obtained in stress relaxation studies of poly-methylmethacrylate as a function of temperature are illustrated in Figure 7-1.   There have been several successful devices used to bring these data together.   For example, a modulus tempera-ture master curve based on the data for constant times can be used, Figure 7-2.

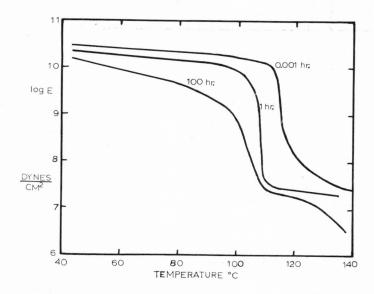

Fig. 7-2. Modulus-temperature master curve based on cross sections of Figure 7-1.

The possibility that joining the various curves end to end would yield a continuous curve immediately suggests itself and this can be done as shown in Figure 7-3.
The transposition to make such a change is the use of a shift factor a_T calculated relative to a chosen reference temperature T_R which can be the glass transition temperature

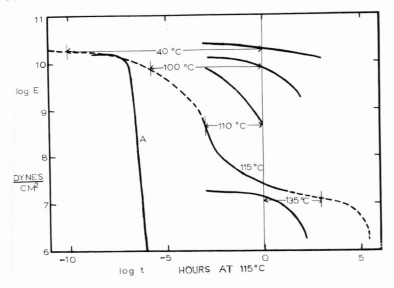

Fig. 7-3. Modulus–time master curve based on time-temperature superposition of data of Figure 7-1. Times referred to temperature of 115°C.

or another close by. For the example above the reference temperature T_R is 115° whereas T_g is 105°C. If the shift had been calculated relative to T_g, a_T would be expressed by

$$\log a_T = \frac{\log t(T)}{t(T_g)} = \frac{-17.44(T-T_g)}{51.6 + (T-T_g)}$$

in which T_g is the chosen reference temperature (T_R). The resultant curve expresses the relaxation modulus over more than 15 decades of time. Of equal value is the use of frequency in place of time. Obviously a long time is a small frequency of application of stress and relaxation and corresponds to results obtained at higher temperatures at a shorter time whereas a short time can be equated to high frequencies which relates to results obtained at lower temperatures over longer periods of time. The practical importance of the superposition principle cannot be overemphasized and more use of the consequences should be possible in practice. However,

care must be taken to recall the assumptions made, an amorphous
polymer studied near its glass transition temperature. In the
previous chapter the principle was applied to elastic liquids.
It has also been applied to blends of polymers [53].

The above is based on the implication that polymers are
uniform and well-behaved. In fact, the actual response yields
a distribution of relaxation times, and what is observed is a
broad peak, the center of which is called the characteristic
relaxation time for convenience. The significance is largely
theoretical once the existence of such a distribution is recognized and its importance noted. The error which arises by
assuming only one Maxwell element is evident from comparing
curve A of Figure 7-3 with the whole master curve.

Effects of Variables on the Modulus

A discussion of viscoelasticity would best be closed with
an illustration of the effect of some of the variables discussed earlier on the modulus as a response. For example, as
the molecular weight increases and as the cross link density

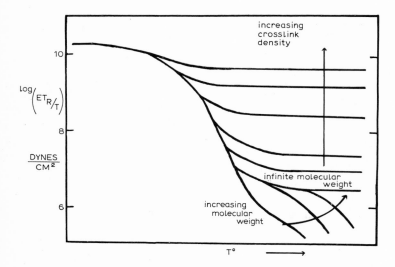

Fig. 7-4. Qualitative effects of increasing molecular weight
and cross-linking of the master curve (after 5).

96

increases the modulus corrected for temperature by use of the reference temperature and plotted against the temperature yields a series of curves, Figure 7-4.

As the molecular weight increases the modulus at higher temperatures increases, a fact of importance in making load-sustaining products. Likewise, as the density of crosslinks increases, not only does the modulus increase as expected, but the product behaves in an increasingly thermoset manner, i.e. the modulus decreases but slightly with increasing temperature. To the left of the envelope line the polymer is glassy.

Somewhat similar results are obtained with semi-crystalline polymers if one considers that crystallites behave like crosslinks. Since these will disappear at the melting point, there will usually be a sharp decline in the modulus at that temperature. A more striking effect if illustrated by water acting as a plasticizer in the amorphous region of nylon and dramatically decreasing the glass transition temperature and, to a lesser extent, the modulus, both above and below the glass transition temperature, Figure 7-5.

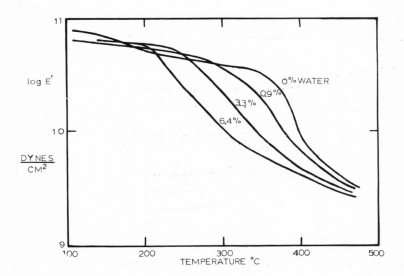

Fig. 7-5. Dynamic modulus vs. temperature for nylon 66 containing various amounts of water (after [5,54]).

An interesting series of data shows the effect of the
chain length of a homologous series of terephthalate polyesters, Figure 7-6 [55], using this technique.

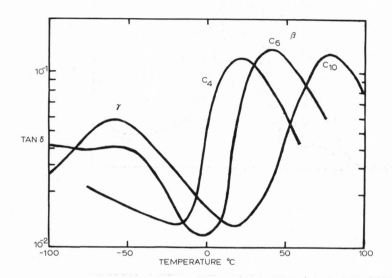

Fig. 7-6. Tan δ versus temperature for terephthalate polyesters based on tetramethylene, hexamethylene and decamethy-
lene glycols.

Another measure of modulus, or stiffness, is the logarithmic
decrement, the fraction of energy dissipated per cycle under
free oscillation of a loaded sample, i.e. torsion pendulum for
example. As the sample is heated, it passes through various
transitions at which the energy absorbed increases and decreases. A series of experiments which illustrate the effects of
degree of crystallinity and of crystal-crystal transitions is
summarized in Figure 7-7.

As the crystallinity increases, the peak due to the glass
transition decreases at 400 K. At about 292 K there is a
crystal-crystal transition from triclinic to disordered hexagonal with a repeat unit of 15 rather than 13 chain atoms. At
176 K there is a gamma-transition related to hindered rotation
of small segments. The melting point would be about 600 K and
the last few points on the curves support this.

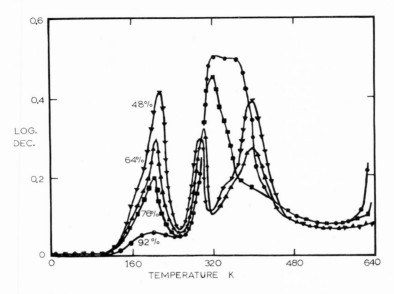

Fig. 7-7. Variation with temperature of the logarithmic decrement for samples of polytetrafluoroethylene of various degrees
of crystallinity (after [56]). Symbols on curves are for identification purposes and are not experimental data.

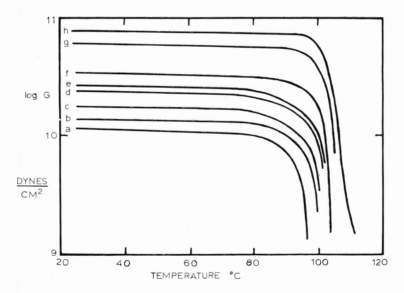

Fig. 7-8. Shear modulus vs. temperature for (a) polystyrene and
for polystyrene containing (b) 20% calcium carbonate, (c) 20%
asbestos, (d) 20% mica, (e) 40% asbestos, (f) 60% asbestos,
(g) 40% mica, and (h) 60% mica (after [57]).

Fillers, particularly reinforcing fillers, act as cross-linking points also. In a three dimensional plot of damping versus filler loading and temperature, it decreases with increasing temperature and with decreasing loading of filler, although not in a regular fashion.

As one would expect, the presence of fillers raises the modulus and to a lesser extent the glass transition temperature of a glassy polymer, such as polystyrene, Figure 7-8. A further discussion of composite and filled systems will be found in the next chapter with emphasis on ultimate properties.

The importance of a good understanding of viscoelasticity cannot be over-emphasized. The more extensive treatments to be found in the references are recommended for additional study. A familiarity with elastic liquids and viscoelasticity is essential for an understanding of the following chapters.

8

Ultimate Properties, Failure Processes

The uses to which a polymer are put require prior knowledge
of the properties and strength of that polymer. The rheological and viscoelastic properties have already been described,
but only in the context of a property at small deformations,
not as a limiting behavior. For example, the tensile strength
of a polymer would be the stress at or near failure and is
called the ultimate tensile [5,23,58-60]. The form of the stress-
strain curve may vary greatly, reflecting the characteristics
of the polymer. The curve may be steep with failure at a high
stress and low strain. The converse may be true, failure at a
low stress and high strain. Then there may be yielding by
which the polymer sample does not fail, but rather draws or
yields and strain may increase greatly at more or less constant
stress. The product of stress and strain is work done so that
the area under the stress strain curve is a measure of the
energy to failure, or toughness. If the sample is not strained
to failure, it may retract elastically, yielding a retraction
curve often below the extension curve; the difference between
the areas under the two curves being the hysteresis or energy
lost during the cycle. If the strain is plastic, the energy
appears as heat and the sample is permanently strained. Intermediate cases will be described later, as will also the fact
that there is a distribution of stresses or strains to failure
depending upon the sample. For this latter reason, the ultimate properties of a material are sample properties.

On the other hand, the modulus, the ratio of the stress to
strain at low values is a material property and hence is more
frequently quoted as a number representing the possible perfor-
mance of the material. Since strain for a given stress is a
function of temperature, the test temperature must be kept
constant and specified. Frequently, the portion of the stress

strain curve, yielding a linear portion obeying Hooke's Law is used to calculate the stiffness, the stress or load which a sample will bear without permanent deformation and the area under this portion of the curve which will be the same both during increasing and decreasing stress is the resilience. The change of the value of the stiffness with temperature is an important item of information in practice.

Constant Rate of Strain

Ideally, tests should simulate the use to which a polymer is to be put, but the uses are so varied and complex that reliance is placed on a few relatively simple, but revealing, measurements and the final approval, or disapproval, made by customer evaluations. One of the earliest and still most use-ful tests is the constant rate of strain tensile test. This has many variations, only a few of which will be described. It should be noted that the strength observed is far below theore-tical, which suggests stress concentration along an advancing front during failure.

The sample most used is dumbbell-shaped with the broad ends clamped in jaws in an instrument which separates them at a constant rate. Failure will be in the thin center portion. Since the sample elongates, some means of measuring the strain must be provided. This may vary from measuring the elongation of the thin portion between two marks on the sample by a ruler, noting the length at failure, to the use of extensometers which automatically record this distance. If ring specimens are used and pulled apart by two "frictionless" pulleys the strain and time are equivalent eliminating the need for the extensometer but there are other errors introduced, such as distortion of the ring and stress concentration at the pulleys. Fibers can be tested similarly, but local necking or yielding at the jaws may introduce errors which can be minimized by using a very long sample or by multiple strands around spools similar to the ring test.

Tearing is another test which is important in practice, but difficult to conduct satisfactorily. It can easily be demon-strated by hand that initiation of a tear may be much more

difficult than the propogation of the tear. Initiation of
failures, including tear, will be discussed later, but, for now,
the usual technique for obtaining useful data is to notch the
center of a crescent-shaped sample, or to use a so-called
trousers tear sample. In either case, the concentration of
stress and its direction at the point of tear may vary, but
repeatability is possible.

For glassy and thermoset samples, the flexural test may be
used in which the beam is held at one, or both, ends and bent
to failure. The most common test uses a simple beam held at
one end and a load is applied by a sliding support. For adhesives, the pull on a flexible adhesive backing, or substrate,
is measured in a manner similar to the trousers tear test, the
peel test. Finally, compressive strength may be measured, such
as the crushing strength of a rigid foam or the deformation
and recovery of a flexible foam, which may be partly or wholly
elastic or plastic.

Breaking Energy

When a suitable stress-strain curve is obtainable, the
breaking energy can be calculated. However, it may be more
convenient, particularly for tough, plastic materials, to apply
more than enough energy to cause rupture and measure the unexpended energy as, for example, by an impact test in tension or
flexure. The latter is represented by the Izod and Charpy
impact testing instruments. The Izod test utilizes a notched
cantilever sample struck by a pendulum hammer and the height
to which the hammer rises after breaking the specimen is a
measure of the unexpended energy; the difference between this
and the initial potential energy being the impact strength
when calculated for a standard sample size. The Charpy test
is similar except that the sample is held at both ends and is
not notched. Rubbery or leathery materials will not break
under such tests, but film samples may be tested using a burst
test accomplished with a falling dart or ball if, for some
reason, the normal tensile test is not satisfactory. However,
it is important to know when a rubbery or leathery material
would become subject to brittle failure with reduction of tem-

perature. This can usually be estimated by measuring the
melting point of semi-crystalline materials or the glassy transition temperature but, a brittleness test is available whereby
the sample is subjected to repeated impact as the temperature
is reduced until fracture occurs to give the brittleness temperature. There are many other tests which could be described
but those quoted illustrate the main principles.

Creep Failure

The tests described above have been a once-through and
comparatively quick with fixed conditions, except for the
brittleness temperature, for which the temperature must be
varied. In some ways it represents a transition test to the
creep failure types in which the stress is applied for a long
time, combined with a temperature change. When the strain
increases rapidly, this is an indication that the deflection
temperature has been reached, the temperature at which creep
becomes rapid. This is usually near the glass transition temperature. The test is applied to samples to estimate the upper
service temperature of a product, but more specifically, it
may be applied to water pipe under pressure or stressed samples
exposed to solvents or other aggressive environments to measure
environmental-stress cracking. These subjects will be discussed
later under failure mechanisms.

Fatigue

Instead of changing time and temperature, the frequency of
application of the stress may be changed. Such tests are called
fatigue tests since the sample will often fail after repeated
tests at a stress well below that required for a single application of stress. Also, the number of cycles to failure usually
increases with lower applied stress, but there may be a lower
limit, endurance limit, or fatigue strength below which the
material appears to resist flexing indefinitely. This is illustrated in Figure 8-1.
Some rigid plastics may fail suddenly with a rapid increase
in temperature whereas a rubbery material may fail more slowly

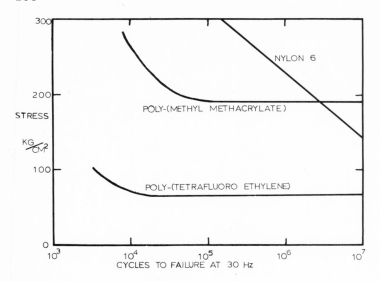

Fig. 8-1. Fatigue life curves for nylon 6, 4.5 mm thick, poly-
(methyl methacrylate) (PMMA) 6.25 mm thick, and polytetrafluoroethylene (PTFE) 6.5 mm thick (after [61]).

with a crack developing and an increase in temperature due to
the hysteresis, which in turn may cause oxidative failure.
More will be said about these phenomena also in the discussion
of failure processes. The purpose of the above is to outline,
in general terms only, what types of tests are applied to
samples representative of the materials being studied and simulating service conditions. Now the problems of relating the
data obtained on a few tests, or within narrow experimental
conditions, to actual service conditions beyond those conditions must be considered.

Distribution of Data

One characteristic, which immediately becomes evident when
the ultimate properties are measured, is that there is a distribution of results. Consequently, to present a single number
to represent the material, an average must be chosen. Later,

the explanation for higher and lower results will be given.
The representative average may be obtained by plotting the
frequency of a given answer against that result, giving a distribution curve, Figure 8-2.

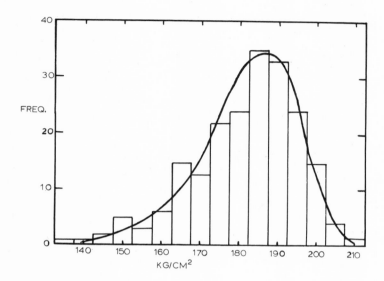

Fig. 8-2. Frequency histogram of tensile strengths of 200
dumbbell specimens from a natural rubber (after [62]).

Alternatively, the response (for example, the stress at
break) may be plotted in descending order of magnitude against
the logarithm of the sample number on doubly exponential paper
and the mode, median, mean and standard deviation calculated
in the usual way (Figure 8-3). Whichever approach is used, a
series of values will be obtained under various experimental
conditions, such as changes in temperature, or rate of strain,
and if the polymer does not undergo some phase changes, such
as crystallization on stretching, the data may be submitted to
a Boltzman type superposition to indicate or predict perfor-
mance under conditions far from the actual experimental conditions. It should be noted here that the failure is a sample
property and the probability of a failure point decreases with

with decreasing sample size. Again, we shall discuss this later.

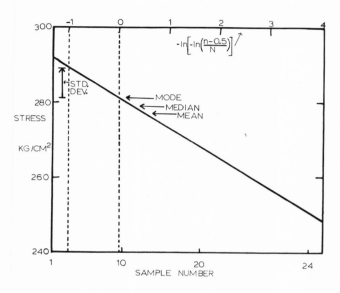

Fig. 8-3. Stress at break plotted for each sample (after [63]).

Reduced Variable Failure Correlations, Failure Envelopes

The stress strain properties depend very much on temperature, rate of strain, the presence of fillers, the occurrence of orientation or crystallization, geometry of the sample, and thermal history. However, for amorphous-non-crystallizing polymers, reduced variable correlations are possible and very valuable. This is accomplished in the same manner as the relaxation modulus-time master curves described earlier. For example, if the stress at break is plotted against the reciprocal of strain rate at various temperatures from -68 to 93° then the curves shifted along the rate axis, a reduced (master) stress at break versus strain rate curve is obtained for the reference temperature chosen, 263°K, Figure 8-4. This temperature is well above the glass transition temperature.

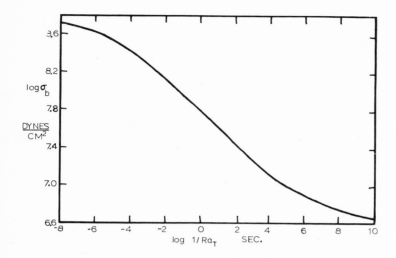

Fig. 8-4. Variation of the tensile strength for a cross-linked styrene-butadiene rubber with reduced strain rate Ra_T. Reference Temperature is 263 K (after[64]).

The shift factor, a_T, is expressed by $\log a_T = -\dfrac{8.86(T - T_s)}{101.6 + T - T_s}$ in which T_s is the reference temperature, in this case 263 K, Figure 8-5. A similar reduced curve may be produced for elongation at break, using the same shift factor, Figure 8-6.

Combining the curves for stress and strain enables a failure envelope to be drawn [65] which eliminates the factors of time and temperature, Figure 8-7. Such a graph obviously has great potential in design applications.

Proceeding upwards on the outside of the envelope represents increasing strain rate or decreasing temperature. If the lines represent constant rate of strain curves, then a time scale may be superimposed on the constant-temperature, constant-rate of-strain lines.

Several other points arise from the figure. The lowest curve is the equilibrium value independent of time. If a sample is stressed to a point D, it will creep with time to F

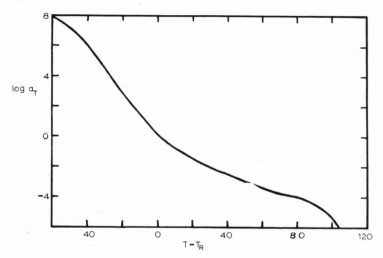

Fig. 8-5. Experimental values of the shift factor a_T from experiment and calculated on the basis of a reference temperature of 263 K, not T_g (after [64]).

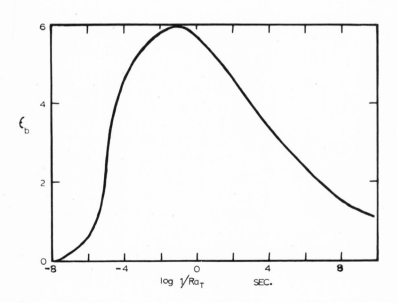

Fig. 8-6. Variation in the strain at break for same samples with reduced strain rate Ra_T (after [64]).

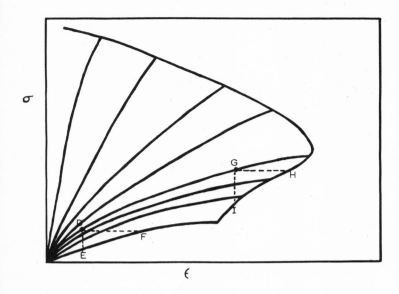

Fig. 8-7. Schematic representation of failure envelope
enclosing curves at constant strain rate. Dotted lines illustrate stress relaxation and creep under different conditions
(after [65]).

or relax with time to E without failure. However, if it is
stressed to G, it will creep to the envelope and fail at H or
stress relax to the envelope and fail at I.

One might reasonably hope that a general failure envelope
for all polymers might emerge. So far this has not been so,
but a compilation of a number of curves has given observation
of remarkably good agreement, Figure 8-8. To take account of
differences in crosslink density, the plot of the product of
strain at break and stress is made against the product of
strain and equilibrium modulus. The data for natural rubber
deviate at higher stresses due to the self-reinforcing effect
of the oriented crystallites. The effect of increasing cross-
link density is a higher stress for a given strain. Orientation or oriented crystallization have the same effect and
becomes increasingly important with greater strain. Recent
studies of very highly drawn fibers or films have shown
greatly enhanced properties.

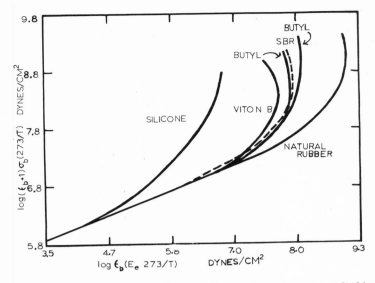

Fig. 8-8. Failure envelopes for vulcanizates of five polymers including Butyl by two curing systems (after[66]).

Fracture of Glassy Materials

When a glassy material is stressed, it will fail at some strain, the value of which varies from sample to sample. Thus, failure of glassy materials, usually brittle failure, is a sample property at first sight. Observation of the fractured sample shows the presence of small voids, small cracks, or crazes which often start at the surface but may be in the bulk of the material. Furthermore, the polymer about the crack is oriented as if it were drawn, Figure 8-9.

This type of failure represents the formation of new surface so that one would expect it to be related to the surface energy and this in turn is related to temperature, Figure 8-10.

The cracks or voids are initiated at some imperfection in the polymer or on the surface; hence, the term "flaw theory" is applied to this type of initiation of failure. The flaw may be impurities, chain ends, or interfaces between crystallites and amorphous regions. It is of interest that as the

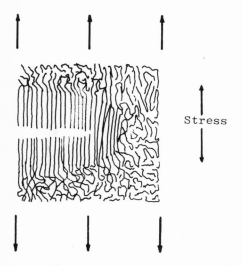

Stress

Fig. 8-9. Molecular orientation about a crack formed during the fracture of a glassy polymer (after [67]).

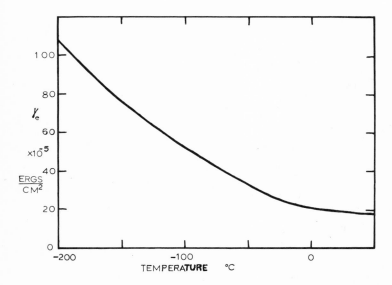

Fig. 8-10. Dependence of the fracture surface energy of poly-(methyl methacrylate) on the temperature of fracture (after [67]).

flaw size is reduced, a limit is reached, the inherent flaw
size, and this is a material property, Figure 8-11.

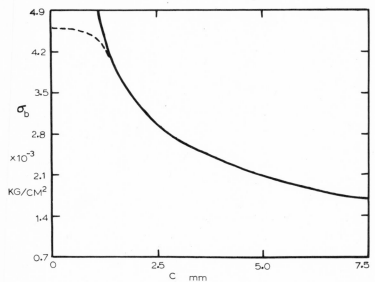

Fig. 8-11. Dependence of tensile strength on crack size in polystyrene samples of various dimensions (after [68]).

For example, the inherent flaw size for poly(methyl methacrylate) is 0.005 cm, whereas it is 0.11 cm for polystyrene.
Likewise, although the failure may be initiated by random
flaws, it is still possible to study the phenomenon. It is
observed that strength is time dependent in that the fracture
stress decreases with the logarithm of the time of stressing,
for example.

It is important to emphasize that during physical testing
standardization of the times and temperatures is necessary
A larger sample is more likely to contain a flaw than a smaller
one, hence micro or small dumbbell specimens of an elastomer
will lead to a high average tensile strength at failure than
larger dumbbells of the same composition. The occurrence of
flaws is random. It is usual to specify whether a macro-
or micro-dumbbell specimen was used. Empirical conversion
factors are available. These must be determined for each
new material tested.

Inhomogeneous Systems, Interface Phenomena

Mention was made earlier of polymeric systems which were not homogeneous; specifically semicrystalline polymers which were a mixture of amorphous and crystalline regions, and polymer-polymer blends held together by adhesion. The subject of mixing of materials is very broad and notably includes polymer-polymer blends, but also polymer-filler blends and polymer-extender blends. We can dismiss the last mentioned with the brief statement that a plasticizer or viscous liquid which is not completely compatible with a polymer may still be dispersed in the polymer as a separate phase. At least as an approximation we can consider this case as similar to a polymer-polymer blend.

When two polymers are blended, both may be amorphous or one may be amorphous and the other semicrystalline or glassy. Alternatively one can consider the crystallites of a polymer as a separate phase from the amorphous matrix of the same polymer. Studies of polymer-polymer blends have been varied and numerous. As has been mentioned earlier a small amount of a rubber in polystyrene enhances the impact strength. At the other end of the scale a crystalline or glassy region in a elastomer will stiffen, harden, or reinforce the amorphous material. However, there must be adhesion between the two phases. If not there may be void formation and failure. Some recent studies interpretated on the basis of changes at the interface between polymers confirms that this is a fruitful field for research [69-72] particularly under dynamic conditions.

However, most activity has centered on the use of non-polymeric fillers and reinforcing agents such as carbon black, glass fibers, clay, talc, silica, glass spheres, and metal powders. These can be divided roughly into reinforcing fillers such as carbon black which enhances some desired property such as the strength and durability of rubber and non-reinforcing fillers such as clay which acts more as a diluent or filler for mastics, sealants, and other inexpensive items not required to withstand high stresses.

The subject of reinforcement by carbon black has been reviewed extensively [73]. Composites generally have also

received considerable attention in the literature [74-76]. Again
one must consider the interface. Reinforcing fillers such as
carbon black have small particle sizes and active surfaces
necessary for adequate bonding to polymers such as the general
purpose elastomers. Natural rubber is self-reinforcing in the
sense that, at high strains, orientation and crystallization
takes place giving strength to the vulcanizate. However, carbon black, while it may not enhance the tensile strength very
much does reduce the strain at break, increases the modulus,
and decreases the rebound which is a measure of the energy
stored and released, i.e. it increases the hysteresis at the
expense of the resilience. On the other hand synthetic elastomers such as styrene-butadiene rubber have poor properties
without carbon black so that in addition to acting as above for
natural rubber, carbon blacks also greatly increase the tensile
strength at failure, five to ten-fold perhaps.

Non-reinforcing fillers may not be as useful because the
particle sizes are too large and the stresses tend to be concentrated at fewer points rather than distributed over many
particles but perhaps more commonly due to a lack of a bonding
to the polymer, a bonding made possible for carbon black by
the reactive groups on the surface. Reactive groups may be
put on silica, ammonium perchlorate or other non-reinforcing
types of fillers by treatment with coupling agents the best
known of which are silanes or silanols. The mechanisms by which
they work may be the subject of controversy but at least the
general idea can be visualized by considering that the silane
derivative bonds to both the surface of the silica [77] and to
the surrounding polymer, Figure 8-12. It is not so easy to see
how this happens with fillers such as ammonium perchlorate [78]
but it may be that water removal from the interface is helpful
and the silane derivatives yield an autophobic surface the
outer layer of which may be wetted by the polymer adequately
for adhesion. Under suitable conditions high energy radiations
will result in bonding of phases, a reaction similar to grafting mentioned earlier. However, often the use of shear will
result in the formation of bonds between fillers and polymers.
This is particularly important in the reinforcement of rubber
with carbon black.

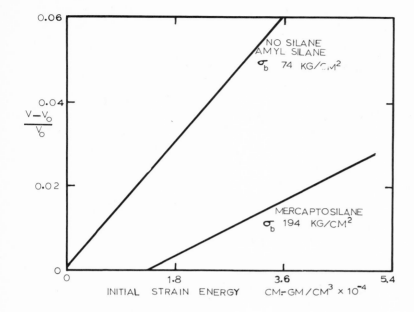

Fig. 8-12. Change in volume on stretching for silica-filled, peroxide-cross-linked ethylene-propylene ter-polymer.

The Mullin's Effect

These inhomogeneous systems, particularly the reinforcing ones which owe their strength to crystallites or carbon black, exhibit a phenomenon known as the Mullin's effect [79]. The first time that a sample is stressed the modulus is higher and as the stress is repeated in a cyclic fashion the modulus decreases to an equilibrium value. A similar effect is observed in the thermoplastic block copolymer elastomers [80] and probably in elastomeric semicrystalline polymers. The explanation is that on the first stressing entangled and dis- ordered chains must be pulled apart to accommodate the move- ment of the crystalline or filler regions to which they are joined. Upon relaxation of the stress the system returns to its original dimensions but the initial built-in entanglements are not restored. Thus on the second and subsequent cycles this extra work of disentanglement does not have to be repeated. Of course if the strain is increased a new set of conditions is set up and the whole process will be repeated. In a recent study, broadening of the damping (tan δ) peaks of systems containing a glassy and an amorphous polymer was observed and attributed to broader distributions of the relaxation times of the polymer at the interfaces. Also the appearance of a third peak between the two was attributed to the equivalent of new phase which could be considered a solid solution of the two components [72,81].

With only this brief guide to composites, heterophase systems, or inhomogeneous systems, and with the implication that interface properties are important, this broad subject must be left to a separate and specialized study.

9

General Use-Related Properties

Ultimate properties of polymers are useful for design purposes since they are the limits of endurance of the polymeric
material. However, in practice, one would avoid destroying
the article so that the stresses and strains imposed are
usually much less than that required for failure, fatigue conditions are avoided, and the article is protected from
environmental hazards. There are a number of properties and
tests which are not destructive which are used to characterize
a polymeric material for various uses. Design criteria can be
numerous for some specialized application, or simple for some
general purpose, non-critical use. A most helpful compendium
of information is the Modern Plastics Encyclopedia where test
methods and data are listed annually.

There may be instances in which pure polymers are used, but
they are rare. Thermoplastic polymers may be mixed with
stabilizers against light and heat, pigments, fillers, plasticizers, lubricants, anti-static agents, fungicides, bacterio-
stats, and with other ingredients for special purposes. Thermosets (vulcanized, cured, cross-linked, hardened) may be likewise compounded but they must also include either a mechanism
for spontaneous curing or added curatives. In both cases
compounding is necessary when the final product is something
as complex as an adhesive, a coating, or a composite.

The simplest tests are those of appearance [82], hardness [83]
and density, which are often the only necessary quality con-
trols. Then there may be mechanical tests which have been
described in the preceding chapters such as rheological, viscoelastic and ultimate properties. Now, we can add to this
list some use-related properties such as thermal, electrical,
optical and acoustical and later, discuss resistance to the
degradative processes of heat and chemicals.

Hardness

Hardness [82] is one of the simplest tests used and frequently
the only one. It is measured by the penetration of a probe
into the surface, a measure of compression modulus or strain
under a fixed stress. Fillers increase hardness, plasticizers
reduce it. In the case of harder materials, such as surface
coatings, the pencil scratch test, based on pencils of various
H numbers, is often adequate and is certainly inexpensive
since pencils are available so readily. More sophisticated
standardized scratch equipment based on the pencil technique
is available.

Density

Density is another property which is easy to measure and is
important in uses where weight is a factor. In general, the
density of a mixture can be calculated from the volume fractions of the various components if the individual densities
are known. However, density may be increased by pressure as
free volume is reduced and the volume approaches the equilibrium value more quickly. Pressure may induce crystallization
and, indeed, the specific volume is a measure of the changes
which take place on crystallization and at glassy transitions
when the temperature is changed. One can describe the compressibility of polymers in much the same way as other materials,
the change in volume relative to a reference volume, say, at
one atmosphere and $25^{\circ}C$.

Thermal Properties

Much of the preceding has stressed the effect of heat on
polymer characteristics, such as the occurrence of transitions
(DSC differential scanning calorimetry and DTA differential
thermal analysis), distortion, deflection, yield, creep, flow,
stress relaxation, stress-strain behavior and ultimate properties, (TMA thermal mechanical analysis), and a later chapter
will cover the effect of heat as a degradative agent (TGA
thermogravimetric analysis and TEA thermoevolution analysis).

There are thermal properties [5, 84-86] which polymers possess
which are characteristic of these as materials. The first is
the coefficient of linear expansion, which is approximately 4
to 40 x 10^{-5} cm/cm $^{\circ}$C as compared with values of 2 x 10^{-5} cm/cm
$^{\circ}$C or less for a steel or other non-organic material. The
symbol often used is α_e and varies with the state of the polymer
and the presence of compounding ingredients. The presence of
fillers may reduce the coefficient so that the difference
between that of the polymer and steel is small enough to avoid
separation or buckling when the two are together and
exposed to a wide temperature range. The thermal conductivity
(k_c) is about 1 to 10 x 10^{-4} cal/sec. cm. $^{\circ}$C; 10^{-2} to 10^{-5}
times less than for metals. It may be increased by incorpora-
ting a metal powder or fiber such as aluminum as a filler.
Conversely, foaming to make insulation decreases the thermal
conductivity to nearly the low value for air, about 0.8 x 10^{-4}
cal/sec. cm. $^{\circ}$C, since the polymer makes little difference.
Finally, the specific heat is typically 0.4 ± 0.1 cal. The
values are slightly higher for amorphous regions of polymers
than for the crystalline regions, and vary widely with the
addition of compounding ingredients being usually close to the
average specific heat of the components calculated from the
volume fractions present.

Electrical Properties

The electrical properties of polymers [5,23,87-98] have much
in common with the mechanical properties. They are conveniently
divided into static properties or the equivalent to electrical
charge and direct current [94] and dynamic properties resulting
from alternating current [95]. The most often used definition
is resistivity, usually expressed as the volume resistivity,
ρ_r, the resistance in ohms of a material one centimeter thick
and 1 cm^2 in area, hence the units, ohm centimeters, Ω cm. The
values for typical materials vary greatly, for example from
10^{17} Ω cm for polystyrene to 10^{-5} Ω cm for silver or copper.
Conductivity is the reciprocal of resistivity and is repre-
sented by σ. It has the units of mho/cm or Ω^{-1} cm^{-1}. The
values of conductivity and resistivity can be altered greatly

by fillers. The volume resistivity may be decreased from 10^{14} to 10 Ω cm by the addition of up to 65 parts of carbon black per 100 parts of elastomer [87,88]. The volume resistivity may also be decreased greatly by the addition of silver powder ($\rho_r = 10^{-4} \cdot \Omega$ cm) or other highly conducting materials, by moisture, or by an increase in temperature. Use is made of conducting fillers and hydrophylic coatings to dissipate static charges on carpets, curtains, flooring or upholstery.

Surface resistivity is the resistivity of surface layers of a polymer including absorbed layers such as moisture. It is measured by placing electrodes on the surface and is expressed as ohms per cm^2. The value can be reduced greatly by antistatic agents which result in a conductive layer on the surface. Dielectric strength is similar to tensile strength. It is the voltage at which a standard thickness of sample will fail. The units are volts (V) per mil (0.001 in), or mega-volts/cm, (1MV/cm = 2500 V/mil). Arc resistance is the ability to allow an arc to start between two electrodes and to continue. If the arc causes volatiles to escape, the arc may stop, but if a conductive path of graphite is formed, the arc will continue.

The most useful function is the dielectric constant (ε) or (K_c) which, in the mechanical analogue, would be the stiffness, or bending modulus. It is the ratio of the capacitance of a parallel plate condensor containing the material to its capacitance in vacuum, or more usually, in air. The units of capacitance are farads, C_f, or microfarads, but the ratio will be dimensionless.

While all of the above measurements are affected by the voltage, the temperature, and various conditioning regimes, the most important variable in practice is the frequency, ω, of the electrical potential, i.e., the dynamic system. While even under static conditions electrons and dipoles in the polymeric material are influenced by the potential applied, this effect may become very great when a dynamic system or alternating sinusoidal potential is applied when the polymers are amorphous, and particularly in a melt form.

When sinusoidal charging and discharging is applied to the polymer sample, the dipoles in the material try to follow the applied potential, i.e. to line-up along the applied field.

Such orientation is small for hydrocarbon polymers, but larger for polar compounds such as poly(vinyl chloride). The effect is decreased by an increase in temperature which has a randomizing or disordering effect due to thermal motion. One must keep in mind that the orientation of the dipoles will be greatly affected by the state of the polymer and will be easier above the T_g and T_m. The dielectric constant, as well as decreasing with increasing temperature, shows marked changes at the glassy and melting transition temperatures. Also, the ability of the dipoles to follow the applied potential is similar to the ability of chain segments to follow an applied stress. There is a retardation time and a relaxation time, λ. This relaxation time decreases with increasing plasticizer or other factors which would enable the dipoles to be more mobile. The relaxation time tends to go through a maximum with increasing frequency, ω, since, at zero frequency, the time is zero and at infinite, or very high frequencies, the dipole does not have time to move and hence there is no time for relaxation, $\lambda = 0$ again.

The movement of the dipoles in the polymer may be divided into two types, an orientational movement which leads to a current of orientation (I_o), and a current of dislocation (I_D). In general, we can ignore the current of rotation and current of dislocation and merely represent the complex current in the dielectric by $I* = I' + iI''$. As in the mechanical analogue, the complex current can be represented by two components at an angle ϕ, so that

$$\tan \phi = \frac{I''}{I'} = \frac{\text{current at right angles, out-of-phase}}{\text{current along axis, in-phase}}$$

Since, in the process of charging the capacitor, the current reaches zero at the maximal voltage, and hence maximal capacitance value, the dielectric constant is 90 degrees out-of-phase with the current. It is convenient to define an angle $\delta = 90 - \phi$. Thus

$$\tan \delta = \cot \phi = \frac{K''}{K'} \quad \frac{\text{dielectric loss (out of phase)}}{\text{dielectric constant (in phase)}}$$

The term dissipation factor (or power factor) is used for

tan δ and it is a measure of the energy dissipated per cycle
or hysteresis loss. The actual power absorbed per cycle, K'',
the dielectric loss index, is the product of the dielectric
constant and dissipation factor, K' tan δ. This is important
since, in a good dielectric, or insulator, the dissipation
factor is low and little energy is lost; whereas, in a poor
insulator, or dielectric, the dissipation factor is large and
the energy loss and heat generation is greater. Likewise,
attempts to heat a polymer with a low dissipation factor
dielectrically will be futile; whereas, heating a polar polymer
with a high dissipation factor is quite feasible. Uniform
heating and/or drying of polar polymers may be accomplished in
this way, including the uniform vulcanization of elastomers
and the uniform heat-sealing of plastics.

If the process of disturbing the dipoles is continued too
long, dynamic fatigue may result, either dielectric failure in
which the polymer is heated to destruction, or treeing in which
paths of irregular shape are literally cut through the polymer.

Aside from the use of polymers as insulators as described
above, polymers with semiconducting, conducting and even
superconducting properties are possible. If the volume resistivity is in the range of 10^{-3} to 10^5 Ω cm, the material can
be considered a semiconductor, if about 10^{-5} Ω cm, it is a
conductor and if it is about 10^{-20} Ω cm it is a superconductor.
For an insulator the energy levels are sufficiently far apart
that the electrons cannot move from one to the other. In
semiconductors, there is the possibility of some electrons
making the transition, usually with the aid of electron rich
or deficient impurity inclusions, while in conductors and
superconductors, the movement of electrons is easy. We do not
propose to discuss in detail the construction and functioning
of inorganic semiconductors. In organic semiconductors and
conductors, the electron orbitals must overlap in some way and
the most usual system is conjugated double bonds -c=c-c=c-c=c-.
If the sequences of double bonds are short as in polyphenyls
and polyacetylenes, the volume resistivity is high, 10^{10} to
10^{15} Ω/cm and the conjugation is called rubiconjugation. If
on the other hand, the conjugated system is more extensive, it
is called ekaconjugation and the volume resistivities of pyro-

polymers and graphites may range from 10^4 to 10^{-4}.

A large number of experimental polymers have been prepared which have semiconducting properties [89]. One of the better group contains a compound tetracyano-p-quinodimethane and it is of interest that the conductivity is greatest in the direction perpendicular to the quinodimethane ring. At about room temperature, this compound, complexed with poly-1-methyl -2-vinylpridinium iodide, has a volume resistivity of about 10^4 Ω cm.

Another type of polymer is a chelated complex including copper atoms [89]. These complexes, the structures of which will not be included here, are stated to have a volume resistivity of about 10^5 Ω cm at 295°C and, surprisingly, about 10^{-20} Ω cm, i.e., superconducting properties at 0°C. Obviously, this is important, but the need for high pressures during formation and crystallinity for superconductivity has so far precluded any known practical uses. A comparative study of electrical properties with dynamic mechanical thermal, and tensile measurements [100] suggested the need for extensive research in this field.

Aside from the complexes based of tetracyanoquino-dimethane and chelate complexes containing metallic atoms, some of the main classes of semiconducting polymers [89] include: poly-phthalocyanines, quinone-semiquinone complexes, polybenzimidazole types (at high pressures), polyazophenylenes, poly-Schiff bases, and various biological polymers. As yet, super-conducting polymers are very much more in the experimental state [99]. A related field is that of magnetic properties of polymers [89,98]. Regularly spaced ferrocene units in the backbone yielded ferromagnetic polymers [101] for example.

Optical Properties

Polymer molecules possess the usual optical properties of absorption spectra, [102,103], luminescence [23], and refractive index [103]. It is the implications of the last mentioned which we shall discuss in more detail, firstly under static conditions and then under dynamic conditions.

There is increasing interest in replacements for glass or

a transparent material for decorative purposes, and there is
more need for controlled transparency for control of illumina-
tion. Polymers, like other materials have indices of refraction (n) which are characteristic of the polymer and a useful
means of identification.

Transparency is defined as the undeviated light $\phi_{undev.}$ =
$\phi - \phi_a - \phi_\alpha$. Transmission would be the light transmitted
($\phi_{undev.}$). The ratio of reflected light ϕ_a to the incident
light (ϕ) yields a reflectance coefficient (σ) and the ratio
of the scattered light (ϕ_α) to the incident light (ϕ) yields
the absorption coefficient, K.

The transmission factor T = $\dfrac{\phi_{undev.}}{\phi}$ is then the ratio of

the transmitted light to the incident light and the attenuation coefficient is the sum of K + σ, i.e. a measure of the
light lost. Since these terms are not specific for polymers,
we can dismiss this subject by simply stating that for optical
uses plastics with suitable coefficients are available for
glazing, lenses and other transparent objects.

It is of more interest to enlarge on the interaction of
light with polymers, in particular, under the titles of optical
rotation and birefringence. An exact understanding of the
ability of polymers to alter the characteristics of light
passing through it is not needed. The incident beam results
in dislocations of electrons in certain bonds or groups of
bonds as, for example, in the plane of a benzene ring. The
bonds are affected, not the center, but the center is usually
asymmetrical, or, if symmetrical, it may be perturbed into an
asymmetrical form, such as a helix. These centers are called
chromophores or optically active centers, but it should be
stressed that the optical activity resides in the bonds, not
the center.

If plane polarized light is used, it may be considered as
composed of two oppositely rotating equal components so that
upon passing through a vacuum or material which is not optically active, the emergent beams appear as a plane polarized
vector. However, if one of the circular waves moves more
slowly, then the emergent plane beam will be rotated. If the
component rotating to the right or clockwise is faster, then

the emergent beam will be rotated to the right, i.e., the
substance through which it has passed is dextrorotatory and the
angle is positive. Naturally the opposite is true for the case
when the component rotating to the left or counterclockwise is
faster; the emergent beam is rotated to the left, i.e. laevo-
rotatory or negative. Since the speed of light in a material
is measured by the index of refraction, the indices of refraction of the two circular beams must differ and so the name
birefringent or birefringence is used. Since mathematically
circular waves are involved, this term is often qualified as
"circularly" birefringent.

If the rotation is a function of the wavelength of the incident beam this is called optical rotatory dispersion [104,105].
The equipment used for such measurements is standard and is
used a great deal in studies of biological materials. If the
two components of plane polarized light are absorbed differently,
the emergent beam is no longer plane polarized, but is elliptically polarized, the medium shows circular dichroism [105].
To measure this requires specialized equipment which often
accompanies the optical rotatory dispersion equipment since
both values may be useful together.

Indeed, the Cotton effect is when both circular dichroism
and optical rotatory dispersion are observed, i.e. both unequal
velocity of transmission of the two components and unequal
absorption, as a function of the wavelength. The interpretation of such spectra is not easy, but considerable use has
been made of the techniques in the study of biological molecules, particularly proteins, but much less use so far in the
study of synthetic polymers.

Since the optical properties discussed are related to the
molecular bonds one can alter the bonds by an applied external
electrical field. The birefringence of the oriented dipoles
is different from that of the random dipoles. This Kerr
effect during which birefringence is studied in an electrical
field is often combined with light scattering measurements
during which depolarization by the molecules in solution is
measured as well as the refractive index increments or polarizability in solution. Since these measurements are largely of
theoretical interest, no further discussion will be included
here.

Birefringence [106,107] may be measured under static conditions as for optically active compounds or polymer crystals,
but it is of more interest when one uses techniques to measure
the birefringence under dynamic or stressed conditions, either
stress-strain birefringence of solid polymer, the photoelastic
effect, or streaming birefringence of polymer solutions.

Studies of dynamic birefringence, stress-birefringence or
strain birefringence are based on the observation that when a
sample of polymer is stressed, the angle of rotation of plane
polarized light passing through the sample increases generally
in proportion to the stress or strain. If the system is semicrystalline, the spherulites are usually not affected at first
and the amorphous phase orients with time, a retardation time
being observed since the system is viscous, and then, if the
stress is held or relaxed, the birefringence decreases again,
as, for example, with molten polystyrene at 98.5°C stressed
over a six second time period, Figure 9-1. The modulus and
birefringence data follow the same general curves during
relaxation.

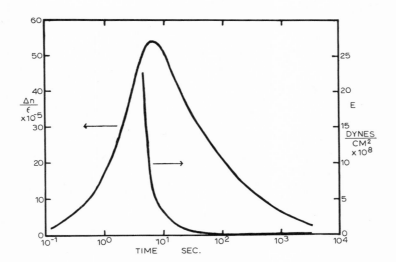

Fig. 9-1. Stress relaxation and birefringence change for two
samples of polystyrene that were stretched slowly (6 sec).
Note that the birefringence can be followed right through the
stretching (after [108]).

With proper instrumentation, the relaxation times can be
measured, but, as mentioned above, the system may be complica-
ted. If the two components are in-phase, plane polarized light
emerges, but if they are out-of-phase, elliptically, or circularly polarized light, emerges. Then, as the stress, or
strain, increases, the spherulites and crystallites are deformed adding to the birefringence.

It is well to recall, at this stage, that this technique
assists in calculating the first $(P_{11} - P_{22})$ and second
$(P_{22} - P_{33})$ normal stress differences since the differences
in the refractive indices (n_{11} in direction of stress, n_{22}
perpendicular to stress, n_{33} lengthwise in sample) $n_{11} - n_{22}$
and $n_{11} - n_{33}$ can be used to calculate $P_{11} - P_{22}$, $P_{11} - P_{33}$,
and $P_{22} - P_{33}$. The route to these values are plots of Δn or
$(n_1 - n_2)$ versus stress or strain, yielding, hopefully, a
linear plot versus shear in the appropriate direction. The
three directions in a typical rotating cone and plate type
instrument could be tangential, axial, and radial.

The calculation of the first and second normal stress differences may not be required very often, but the method may be
used with success to study the properties of elastic materials.
Data for butadiene-styrene copolymers are in Figure 9-2 as
a function of comonomer content.

Note that as the styrene content increases, the birefringence divided by stress decreases and the peak shifts towards
higher temperatures, there is less flow and less orientation
at a given temperature and the peak is reached at the glass
transition temperature. An exception is the 5/95 styrene butadiene copolymer, which is partially crystalline and, hence, has
a melting peak at about room temperature and a higher value for
the birefringence divided by stress, since the remainder of the
polymer is amorphous. There is some evidence of a glass transition at low temperatures.

The values for birefringence of typical polymers may be
found in various tables. A few values are polyisoprene 1,949,
2,200, polybutadiene 3,200, polystyrene at 278K 10.2 (glassy),
polystyrene at 388K −5,200 (molten), all $\Delta n \times 10^5$.

Still another use for the technique is to use the numerical
data as a means of relating the composition to the measured

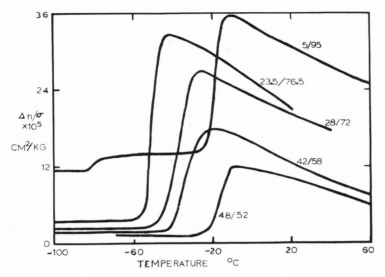

Fig. 9-2. Birefringence per unit stress of styrene-butadiene copolymers as a function of temperature (after [109]).

values to determine which phase of a polyblend or block copolymer is undergoing strain. For example, an isoprene block copolymer with styrene yielded the data in Figure 9-3.

At first the polystyrene phase is yielding and taking the strain; hence, a rapid rise in stress, but a small change in the birefringence. Then the polybutadiene starts to be deformed and stress and birefringence increase. Finally, the glassy polystyrene phase disintegrates at higher stresses and the negative contribution of the "molten" polystyrene causes the birefringence to level off although the stress is rising rapidly.

Streaming birefringence [111,112] is observed when a liquid is viscoelastic or a polymer is dissolved and the solution placed under high shear. Some idea of the size, shape and structure of the polymer molecules can be obtained. The molecules tend to line up, polarization is possible, leading to anisotropy, and then there is relaxation. As a polystyrene molecule is stretched out, for example by applying a stress,

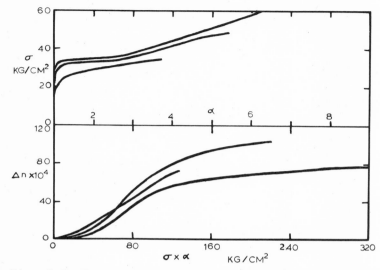

Fig. 9-3. Upper, stress strain properties of styrene-isoprene
block copolymers. Lower, photoelastic properties corresponding
(after[110])

the backbone tends to be oriented and the phenyl rings tend
to point outwards and yield a negative value to the birefringence under shear. Likewise, transpolyisoprene yields a more
positive value than cis polybutadiene, and a higher quantita-
tive result. These measures of polarizability are useful in
estimating the number and strength of dipoles in the structure
of chains and may be used to calculate stiffness of the chains,
entropy changes when molecules go into solution, the effect of
temperature and solvent on the orientation of chains, and to
observe changes of orientation taking place during the stressing, polymerization, crystallization, etc. The results are
usually quoted in Brewsters (C) 10^{-13} cm^2/dyne and some typical
values for polyisobutylene in decalin   1,600, poly(methyl
methacrylate) in benzene   1,000, polyethylene in benzene
2,200, cis-polyisoprene in benzene   2,000, cis-polybutadiene
in toluene   3,500, isotactic polystyrene in bromoform   -10,300,
and atactic polystyrene in bromoform   -6,800 Brewsters   would
illustrate the range of numerical results obtained.

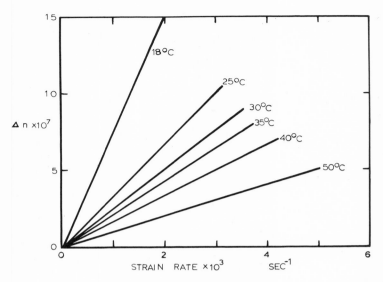

Fig. 9-4. Flow birefringence vs. shear rate (after [111]).

A typical series of data for polystyrene in methyl-4-bromo-phenyl carbinol is in Figure 9-4.

Some practical uses of optical methods must be dismissed with just mere mention. The well-known photoelastic method of stress analysis is based on the birefringence of polymers and aside from its great value in mechanical engineering may be used to study strains in plastic objects and test pieces [112]. Light scattering, particularly of crystallites in amorphous polymers and polymers in solution yields much data on the size and shape of these dispersed phases but the subject is too large for complete discussion, beyond what has been included in earlier chapters. While all of the above has applied to visible light, similar considerations apply to other regions of the electromagnetic spectrum, x-ray, infrared and ultraviolet particularly. Nor can we discuss in detail the advances made possible by polarized monochromatic light from lasers and the combined use of mechanical and optical methods [106]. Various research studies using polarized infrared radiation have been reported but the technique has not found general use in the study of polymers.

Acoustical Properties

In general, the acoustical properties of polymers [114] are
the mechanical properties measured over a sonic range of
frequencies, but there is sufficient interest in noise abatement to warrant a separate discussion. Also, whereas mechanical deformations are caused normally by contact of a driving
mechanism with the sample, this driving mechanism may be more
complex in acoustical studies. The origin of the energy and
the response may be separated, not only by the material under
study, but also by a path of air or water, or other matter
which serves merely as a connector and neither as an origina-
tor of the sound, nor attenuator or absorber of the sound. In
addition, while instrumental detection and analysis of the
sound is easily accomplished, the relationship of this to the
response of the ear must be considered, often with an intervening
medium in addition to the atmosphere.

Firstly, let us consider the sound intensity factor and
sound intensity level. The former is the ratio of the pressure
exerted by the sound wave divided by the pressure exerted by a
just audible noise. This just audible sound is in the range
of one picowatt per square meter, 10^{-12} w/m^2, in comparison
with a jet engine at 25 meters which may exert a pressure 10^{14}
times greater, i.e. 100 watts/square meter, w/m^2. The sound
intensity level is a logarithmic value and is called a bel,
i.e. the noise of the jet engine would be 14 bels. To make it
easier to work with fractions of bels, the more common term is
decibels, or 10 times the bels, dB.

The second level is a pressure measurement on the ear, but
the ear does not respond to pressure uniformly for various
frequencies. Thus, when converting an energy measurement into
the practical term of loudness, the energy measured electroni-
cally must be attenuated by a filter, simulating the response
of the ear. Usually, a filter known as the 'A' filter is used,
so that the correct response of the human ear can be estimated.
The data are then given as dB(A).

To illustrate the effect of frequency, the value of 50dB
at 1000 Hz is defined as 50 Phons. To obtain the dB value
considered to be equally loud at other frequencies values have

been obtained by panel-comparisons on human subjects and the
values decrease with increasing frequency. Likewise, the sub-
jective impression of doubling the sound or noise has been
estimated. Doubling the noise is an increase of one sone.
The noise level at 40 Phons or dB at 1000 Hz is taken as
standard. A doubling of the sound appears to be about equal to
10dB or 10 Phons increase at 1000 Hz, with the corresponding
changes required to convert to Phons at other frequencies.
Another measure is the noise rating number or sound pressure
level. The values are related to the pressure exerted on the
ear by a noise of 85 dB at 1000 Hz and levels above this are
considered dangerous. Corresponding noise pressure ratings
increase with decreasing frequency.

These sound pressure levels in Newtons/m^2 can be estimated
from the fact that at 1000 Hz the threshold of hearing is equal
to a pressure of about 2×10^{-5} N/m^2. The sound intensity
level is the sound pressure squared, i.e. watts/m^2, approximately 10^{-12} W/m^2 as stated earlier.

In practice we are concerned with a spectrum of sounds and
to simplify the measurements, the values are usually determined at a number of octaves and added, although automatic
integration over all frequencies is possible. The octaves are
based on 63, 125, 250, 500, 1000, 2000, 4000 and 8000 Hz with
31.5 and 16000 Hz being included at times. The ear can detect
both higher and lower frequencies, but these are seldom a cause
of concern, the very high frequencies having little energy and
the very low frequencies falling more into the mechanical damping study except the very low frequencies of an organ. Of
course, these frequencies do carry a lot of energy.

Although much has been done in the study of acoustics, this
has usually been aimed at scientific aspects or the practical
aspects of tuning concert halls, musical instruments, etc.
With the advent of interest in noise abatement and sound control,
much more of a practical nature has evolved, but the theoretical concepts have yet to be collected together and analyzed.
Thus we shall discuss the remaining aspects in general terms
only.

Recalling the dynamic mechanical properties of polymers, it
is not surprising that there are ranges over which a polymer

will absorb certain frequencies, usually in the temperature
range between the glassy and the rubbery state. Allusion was
made to this in an earlier chapter on the transitions in polymers.

The practical aspects of sound absorption can be divided
into the control of airborne sound and structurally borne
sound. The first mentioned would be the case of noise from a
telephone or a typewriter. Assuming that there is no control
at the source, there are two approaches, an intervening absorbing system and a non-reflecting surface treatment to prevent
reflection and resonance or reverberation. To measure the
effect of materials in this type of system, we use a standing
wave apparatus at one end of which is produced a sound of known
frequency. This travels the length of the tube, penetrates or
is reflected from the test sample, then is reflected from a
backing plate, and travels back down the tube towards the
source. If there has been absorption or attenuation, then a
standing wave will be set up and this can be found by a probe
microphone moved along the axis of the tube on a rod. This
type of equipment is useful in studying acoustical ceilings
and panels, upholstery, drapery, or flooring. A number of
practical solutions to the problems may be listed.

New typewriters have a plastic cover which is usually poly
(methyl methacrylate). This absorbs considerable noise. A
similar principle is to use fixed panels as in windows but with
energy absorbing seals to the frame. The panel vibrates, but
the energy is absorbed by the seal. More closely allied to
the typewriter cover is the free floating panel which vibrates
or resonates but the sound is absorbed since the energy is
dissipated as heat. Poly(methyl methacrylate) type panels are
often hung in stores and offices.

Textured absorbers are also common as ceiling and wall
panels, carpeting, and textured fabrics including vinyl and
other plastic coated fabrics, glass fiber curtains, and vinyl
curtains filled with lead powder. These may be augmented by
fibrous or foamed materials, glass fiber layers or similar
materials, light weight concrete or plaster, open-cell foams
which trap noise in their pores, closed-cell foams which may
reflect or absorb sound, and open-celled foams which are

134

coated with films to keep them cleaner and to strengthen them. The coating on an acoustical foam may move the sound absorbing characteristics to lower frequencies, probably due to the drumming effect of the film over the foam, i.e. the sound passes through and resonates in the foam structure where the energy of the higher frequency waves is converted to heat by friction at the boundary layers between air and polymer more than is so for the lower frequencies which emerge again from the foam. A recent study [115] using the standing wave equipment would illustrate the type of data which can be obtained, Figure 9-5.

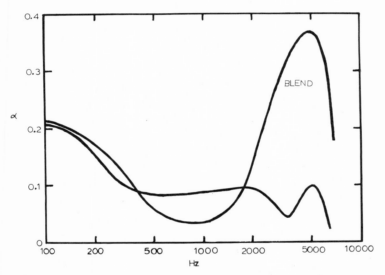

Fig. 9-5. Normal absorption coefficient α of an epoxy resin and of an epoxy resin containing 33 parts per hundred of resin of hydroxy terminated polybutadiene, 27°C, 6 mm thick.

A somewhat different picture pertains to structurally borne sound [116]. This is measured by a damping modulus apparatus in which a strip of the polymer or a strip of metal to which a layer of polymer has been adhered is resonated by force at various frequencies and the magnitude of the modes or resonance peaks if any observed. An example of data obtained is in

Figure 9-6 for an unsupported strip of polymer.

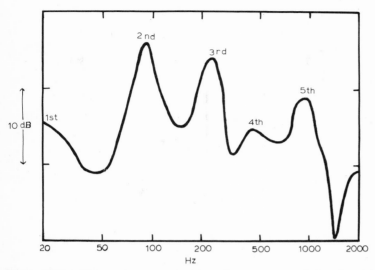

Fig. 9-6. Frequency response of a commercial plasticized poly-
(vinyl chloride), butadiene-acrylonitrile elastomer blend. (Geon
552 solids, Goodrich) (after [117]).

Another technique is the logarithmic decrement under free
oscillation of the sample. This is comparable to deadening the
sound of a rattling piece of metal with a layer of sound
deadening material. The result is much reduced transmission,
reflection or emission of noise from the metal surface. This,
again, finds use in panels, curtains, linings of motor chambers
and vehicles, linings of buildings, and coatings for vibrating
vessels and structures.

One of the most popular methods is to make a foamed polymer
sandwich with lead film or composites with heavy powders such
as lead or barytes, and join these to the vibrating body. The
film or heavy particles resonate, and the energy is absorbed
by the foam or polymer. There are numerous commercial panels
for this use also which consist of an adhesive layer, an energy
absorbing foam layer for structurally borne sound, a lead film
to resonate, another foam layer for absorption of air borne

sound, and a covering layer to keep the whole clean and dry.
This composite may be 5 cm thick. The results obtained with
such a composite are quite good. The most recent developments
omit the lead and use the damping properties of a single layer
of polymer without the need for the resonating metal or
filler [115,117]. Similar in some respects are the silent paints
composed of two layers, a damping layer and a constraining
layer [118].

Other techniques of preventing structurally borne sound
include elastomeric sections in metal ducts and pipes, rubber
mounts or dampers on engines and pumps, or floors floating on
expanded polystyrene or other energy absorbing materials.
Numerous examples of such items will be found in design journals
and include such well-known items as the rubber pads under sub-
way rails, bridge supports or high rise buildings on vibrating
ground.

Some special items are worthy of at least short mention.
The noise of explosions in engines is smoothed out by mufflers.
These systems can also be used on pneumatic drills. There are
silencers which can be built into valves and pipelines which
cut down the noise from these pieces of equipment. If all else
fails these pieces of equipment can be covered with a sound
damping material or enclosed in an airborne-sound absorbing
case.

Similar considerations apply to water-borne sound. The
sound may be absorbed by a textured surface or a surface with
a gradation of properties, in each case having sound absorbing
properties. The design is to avoid reflections, the so-called
ρC (density (ρ) times the velocity of sound (C)) approach of
gradation of properties from the ρC of the water to the ρC of
the substrate. Alternatively, and perhaps more simply, the
damping cover may contain air bubbles or chambers which act as
resonators until the energy is absorbed. A somewhat similar
structure is present in the skin of porpoises, but the holes or
chambers are filled with oil.

A few definitions should suffice finally in this introduction to the subject. Acoustical fatigue can be observed [119].
The repeated impact of the sound causes the temperature to rise
and the material to fail, just as in mechanical fatigue,

usually from thermal failure.

We have neglected ultrasonic properties. As a first approximation, these may be considered to be similar to acoustical, but one should keep in mind that the frequencies may be so high that the polymer cannot follow the waves, the relaxation time is too long, and cavitation will result, with enormous releases of heat. This is useful in ultrasonic welding, but may be catastrophic in uses where the input continues uncontrolled.

Finally a word about very low frequencies, infrasonics [120]. These vibrations from large structures and in automobiles have been little studied but may well be related to the malaise of persons living near such sources, and to car-sickness.

10

Degradation

Degradation is any process [5] by which the original polymer structure is changed and failure results from that physical or chemical change [121,122,123]. It should be differentiated from ultimate failure phenomena such as tensile or viscoelastic failure or abrasive wear. It should also be differentiated from the failures which result from changes in properties due to increased temperature, or to absorption of water or plasticizer which in effect alter the ultimate properties. There is a distinct change in the molecular structure, usually of the molecular weight or composition. However, the distinction between failure phenomena and degradation is not clear-cut. This is illustrated by separation of reinforcing fillers from polymers (adhesive failure) in the presence of moisture.

Degrading agencies are numerous and often more than one may be effective under given conditions such as during weathering or exposure to corrosive environments. There may be the imposition of energy in the form of heat, whether by mechanical action (including ultrasonic and sonic energy), radiation, such as gamma rays, X-rays, visible light, ultraviolet light, or infrared radiation, or electrical, in the form of dielectric effects. Then there are chemical effects such as non-abrasive wear, mastication, oxidation, ozone attack, hydrolysis, chemical attack by solvents and detergents, cracking due to swelling by plasticizer or hardening due to loss of plasticizer, migration of plasticizers from one layer to another, crazing and cracking, delamination or debonding, void formation, biological attack, and chemical degradation resulting in loss of portions of the molecule or in cross linking, cyclization, scission of the chains, and changes in the side chains which bring about any of the above changes.

Thermal

One of the most interesting is thermal depolymerization by heat, the reverse of polymerization. This degradation is started at some weak point along the polymer chain. This might be an -O-O linkage, a chain end, an impurity, or a head-head linkage. This starts the peeling off of monomer units or unzipping of chains. The process is not perfect, but if can be illustrated by typical cases. The usual criterion is the amount of monomer produced per initiation step, although other products may be produced as well. Typical results are in Table 10-1.

Table 10-1
Thermal degradation of polymers[*]

Polymer	Heat of Poly-merization K cal/mol	Weight % of Monomer Produced	Minimum Zip Length
poly(methyl methacrylate)	10-13	>90	>>200
poly(alpha-methylstyrene)	9.5	>90	>>200
polyisobutylene	10	20-50	3.1
polystyrene	17	65	3.1
polybutadiene	16-18	20-30	–
polyisoprene	17-20	20-30	–
polyethylene	22-25	<1	0.01

[*]After [121]

The first extreme can be visualized. When a molecule of poly-(methyl methacrylate) is broken over 200 monomer units will peel off during the unzipping process, and little else. On the other hand, if polyethylene is cleaved, there is only one chance in a hundred that an ethylene unit will be eliminated. Other products may be formed, but the depolymerization process virtually does not exist. In between there are the cases when some monomer is produced and the number of units liberated is small, polystyrene and polybutadiene. Since the depolymerization is the reverse of polymerization it is not surprising that the reversal is easier and more efficient when the heat of polymerization is low. Another criterion which can be applied is that the unzipping process is more probable when the non-backbone bonds are strong as in polyvinyl chloride, polyvinylidine chloride and particularly in the fluoro-polymers such as polytetrafluoroethylene. The importance of the long zip lengths must be noted also. If the degree of polymerization is low then the polymer chain will completely unzip in the case of poly(methyl methacrylate) and one is left with undegraded polymer chains and monomer. There is no change in the average molecular weight of the remaining product. As the degree of polymerization increases, then there will be small fragments left as well as the original long chains and the average molecular weight will be decreased. On the other end of the scale, the cleavage of a polyethylene molecule will leave two fragments totalling the size of the original so that the average molecular weight decreases with time.

The termination reaction which stops the unzipping will vary, but for poly(methyl methacrylate) will likely be disproportionation and for polystyrene mutual recombination, the reverse of the initiation step or similar to the termination step of polymerization.

The decomposition of the polymers may be followed by various techniques which will not be discussed in detail. There is differential scanning calorimetry (DSC) and differential thermal analysis (DTA), which will measure exotherms and endotherms indicating decomposition and the amount of it. There is thermogravimetric analysis (TGA), by which weight loss is measured and this may be coupled with gas chromatographic (GC), mass spectrometry (MS), infrared analysis (IR), ultraviolet analysis (UV), nuclear magnetic resonance (NMR), or thermal evolution analysis (TEA) to measure and identify the gaseous and liquid products. The residual polymer may be studied using intrinsic viscosity η, osmotic pressure (π), sedimentation (S), or gel permeation chromatography (GPC).

Ceiling Temperature

Another useful concept arising from the measurements aside from knowledge on the thermal stability of polymers and the

nature of the products of decomposition is the ceiling temperature [121] at which the rates of the forward reaction yielding
polymer and the reverse reaction yielding monomer are equal.
The temperature at which this takes place is the ceiling temperature. This temperature may be very high or it may be at or
near room temperature. For example, chloral [126] may be kept
molten mixed with catalyst but on cooling below its ceiling
temperature of about 50°C, it will polymerize. This is one of
the monomers which can be used to make castings by polymerization in situ, a growing technology.

There are several other techniques of thermal degradation
which differ somewhat from the above. The first is mechano-
chemical, whereby the energy of mechanical work is used to
break the chains and no heat or oxygen need be applied. The
mechanical work cleaves the chains which may then depolymerize
as above, or recombine. If oxygen is present, the radicals
formed will react with oxygen, as will be discussed later.

Chemical Conversions

Still another type of thermal degradation is dehydrohaloge-
nation whereby hydrogen chloride removed from polyvinyl
chloride leaving an unsaturated structure. The hydrogen
chloride catalyzes further degradation. This structure was
mentioned before as a pyrographite and has semiconducting
properties. The retardation of the dehydrochlorination of
polyvinyl chloride may be accomplished if hydrogen chloride
acceptors, such as calcium, barium, or lead salts, or amines,
are present.

Another type of thermal degradation is really a thermal
polymerization whereby polymers, such as poly-1,2-butadiene or
polyacrylonitrile are internally polymerized to yield ladder
structures. The former may be illustrated as follows:

A somewhat similar process takes place in the cyclization of natural rubber. All three processes are commercial and used to produce resins, usually with exceptionally good resistance to heat, since unsaturation is lost and stable ring structures are formed. The inhibition of thermal reactions [127] is most difficult and there is no general way of stopping them. Obviously, if they can be initiated almost anywhere in the bulk of the polymer, an additive is not helpful. The processes may be free radical or ionic so that free radical inhibitors, such as are used against oxygen are sometimes, but not always, effective. Ionic compounds, such as acids or bases, may be helpful, but, in general, the thermal degradative reaction proceeds without much control beyond that afforded by stronger bonds and a naturally thermally resistant polymer, of which there are now a great many. In those cases in which ultraviolet light is the cause of initiation, ultraviolet absorbers will inhibit the initiation. Likewise, if the less specific initiation by X-rays or gamma rays is the cause the one can sometimes use protective agents against the high energy radiation.

As a final comment one should add that the gases produced by thermal degradation of polymers are fuels for combustion, a subject which will be discussed later. The various types of photodegradable plastics currently coming onto the market can be listed as a special case of thermal degradation. These are designed to absorb light energy (usually ultraviolet) and to degrade into a powder or gas which disappears. We shall not discuss the chemistry of these in detail.

Slow Oxidation and Ozonolysis [121-122,128]

The title of "Slow Oxidation" is chosen to deliberately set this subject aside from combustion. It and its companion subject of ozone degradation are chemical reactions which take place without a flame or charring. Of course, in practice, the two types meet in a grey area and one may lead to the other, i.e. slow oxidation may result in the accumulation of heat which results in gross degradation followed by combustion.

It is convenient to divide slow oxidation into two main types, one leading to cross linking or hardening and the other

to·chain scission or softening. In most cases, both processes
take place, but a striking difference is noted in the elastomers, polybutadiene cross links or hardens whereas polyisoprene
cleaves and softens.

   We are dealing with another chain reaction, initiation,
propagation and termination. Initiation takes place at the
alpha-hydrogen with active oxygen which may be created by any
number of ways, ultraviolet light, heat, or a chemical reaction
with metallic compounds. It will suffice for our purposes to
say that the polymer immersed in an atmosphere rich in oxygen
such as air will eventually come in contact with active oxygen
and lose an H atom to form a free radical.

```
  H H H H                              H H H H
  | | | |                              | | | |
—C-C=C-C— +   ·O-O·        ⟶         —C-C=C-C— +   HOO·
  |     |                              ·     |
  H     H                                    H
```

It will then react with another oxygen molecule to form a
peroxy free radical and this in turn will react with another
unit of a polymer chain to produce a hydroperoxide and another
free radical, a chain transfer type reaction. These two steps
continue alternately for many cycles and represent the propogation step.

```
  H H H H                              H H H H
  | | | |                              | | | |
—C-C=C-C— +   O₂           ⟶         —C-C=C-C—
  ·     |                              |     |
        H                              O     H
                                       |
                                       O
                                       ·
```

```
  H H H H   H H H H                    H H H H   H H H H
  | | | |   | | | |                    | | | |   | | | |
—C-C=C-C— + —C-C=C-C—      ⟶         —C-C=C-C— + -C-C=C-C—
  |     |   |     |                    |     |   ·     |
  O     H   H     H                    O     H         H
  |                                    |
  O                                    O
  ·                                    |
                                       H
```

Termination may be accomplished by combination of the free
radicals in pairs leading to cross links, either carbon to
carbon or the formation of a peroxide. The former are stable
but the latter may be transitory.

$$2 \quad \begin{array}{c} H \ H \ H \ H \\ | \ | \ | \ | \\ -C-C=C-C- \\ \bullet | \\ H \end{array} \qquad \longrightarrow \qquad \begin{array}{c} H \ H \ H \ H_2 \\ | \ | \ | \ | \\ -C-C=C-C- \\ -C-C=C-C- \\ | \ | \ | \ | \\ H \ H \ H \ H_2 \end{array}$$

$$\begin{array}{c} H \ H \ H \ H \\ | \ | \ | \ | \\ -C-C=C-C- \\ \bullet | \\ H \\ | \\ O \\ | \\ O \ H \ H \ H \\ | \ | \ | \ | \\ -C-C=C-C- \\ | | \\ H H \end{array} \qquad \longrightarrow \qquad \begin{array}{c} H \ H \ H \ H \\ | \ | \ | \ | \\ -C-C=C-C- \\ | | \\ O H \\ | \\ O \\ | H \\ | | \\ -C-C=C-C- \\ | \ | \ | \ | \\ H \ H \ H \ H \end{array}$$

Alternatively the process may be stopped by inhibitors or anti-
oxidants which will combine with the radicals at some early
stage, yielding an inactive radical and breaking the chain.
Inhibitors are often aromatic amines or phenolic compounds.

$$\begin{array}{c} H \ H \ H \ H \\ | \ | \ | \ | \\ -C-C=C-C- \\ \bullet | \\ H \end{array} + \ RH \qquad \longrightarrow \qquad \begin{array}{c} H \ H \ H \ H \\ | \ | \ | \ | \\ -C-C=C-C- \\ | | \\ H H \end{array} + \ R\bullet \\ inactive$$

However, the peroxy groups may cleave to form new radicals,
initiating new oxidation chains, hence, the term "branching
reaction" since for each original reaction with oxygen there
may result many propogation reactions, an autocatalytic process:

$$\begin{array}{c} H \ H \ H \ H \\ | \ | \ | \ | \\ -C-C=C-C- \\ | | \\ O H \\ | \\ O \\ | \\ H \end{array} \qquad \longrightarrow \qquad \begin{array}{c} H \ H \ H \ H \\ | \ | \ | \ | \\ -C-C=C-C- \\ | | \\ O H \end{array} + \ HO\bullet$$

$$\begin{array}{c} H \ H \ H \ H \\ | \ | \ | \ | \\ -C-C=C-C- \\ | | \\ O H \\ | \\ O H \\ | | \\ -C-C=C-C- \\ | \ | \ | \ | \\ H \ H \ H \ H \end{array} \qquad \longrightarrow \qquad 2 \quad \begin{array}{c} H \ H \ H \ H \\ | \ | \ | \ | \\ -C-C=C-C- \\ | | \\ O H \\ \bullet \end{array}$$

Finally, the peroxides may cleave to yield aldehydes, alcohols
or acids and thereby leave the chains cut in two, or the
product softened with a lower average molecular weight. This
is what happens with natural rubber under various conditions.

```
     H  H  H  H                              H
     |  |  |  |                              |
   —C—C=C—C—                 ⟶            —COH        alcohol
     |        |                              |
     O        H                              H
     |
     O
     |
     H
```

```
                                             H
                                             |
                                           —C=O        aldehyde
```

```
                                          >C=O        ketone
```

```
                                           —C—OH      acid
                                             ‖
                                             O
```

Ozone Attack

   Quite a different process takes place with ozone. This
reacts with double bonds of elastomers to yield ozonides which
are relatively stable. However, when stress is applied, the
samples cleave at the ozonides in various ways again to yield
acids, aldehydes, alcohols, or ketones. The process is
believed to be one of stress concentration at an advancing
front and oxygen is necessary. Ozone really initiates a
localized oxidation which leads to the massive cleavage under
stress. Again, there are antiozonants which slow this reaction;
such materials as heptylated diphenylamines. Other means of
avoiding ozone degradation are to make the surfaces stress-free
by careful annealing or by the use of a layer of plasticizer,
or by having an internal oil or wax which bleeds to the surface
and protects the surface from ozone. Ozone cracking is particularly noticeable in areas where there is smog since it is a
component of smog. Saturated polymers such as Butyl or ethylene-
propylene elastomers are resistant to ozone attack and in cases
of need these elastomers are used in preference to the unsaturated types.

   The effect of ozone is so specific and rapid that elastomers
are used as extremely sensitive tests for the presence of ozone.

If the concentration is small there will be a few large cracks, if it is large there will be many small ones. Thus the size and number of cracks is a measure of ozone concentration and the test only takes a few seconds.

Rate of Oxidation

Of general interest is the relative rate of oxidation which, as one would expect, increases with the number of alpha hydrogens. Oxygen uptake varies with time, also. Firstly there is initiation and then there is a fairly constant rate of uptake as the initiated chains propogate. The products of the propogation reaction begin to decompose yielding the accelerating or autocatalytic region. Finally, the reaction will slow down as the products are consumed.

Measurement of oxygen uptake is conveniently done in much the same manner and with much the same equipment as used by biochemists. The volume of oxygen absorbed is measured with time for a given weight of polymer. One can express the relative rates of absorption for various elastomers graphically. Natural rubber absorbs oxygen readily whereas the essentially saturated Butyl rubber absorbs very little. The copolymer of styrene and butadiene absorbs at a rate between the two.

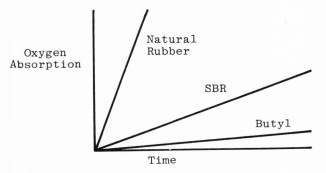

Inhibitors and Retarders

To illustrate the effect of additives, the oxygen absorption of a single polymer, say, natural rubber, can be measured. An inhibitor is able to slow the initiation to the point where

virtually no oxygen is absorbed until the inhibitor is consumed, then oxidation proceeds normally. A retarder interferes with the propogation step, imperfectly, so that the
oxygen uptake is merely slowed. In some cases the inhibitor
gives rise to a retarder, a desirable condition.

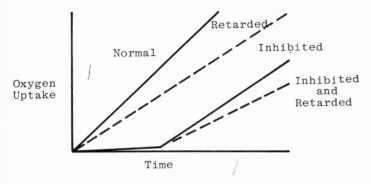

As a special case of slow oxidation, we should list biode-
gradable, since biological systems may decompose plastics by
an aerobic process. We cannot cover all the chemistry
involved. Briefly, bacteria and moulds will metabolize most
plastics which are not protected and the attack can be promoted
or retarded by appropriate compounds, the compositions of
which are often not revealed. Unprotected plastics would last
indefinitely buried in the ground below the levels reached by
bacteria, but on the surface may be decomposed in periods
varying from days to years depending upon the types, the conditions and the pressure of promoters.

Combustion and Flame Retardancy

This is an immense subject, but one which should be discussed at least briefly in any course relating to the engineering aspects of plastics. The reason is that, for many of the
uses, the combustion of the polymer is a factor, its resistance to combustion may be important, and the products of
combustion may be hazardous.

The simplest view of combustion is illustrated by the
laminar flow flame technique which is the combustion of a gas
such as methane flowing in a tube. Firstly the gas is heated

then it ignites and burns with the formation of carbon mono-
xide and water. This flame may have a bluish tinge. Combustion continues with a colorless flame to yield carbon dioxide.
Finally, the flame cools, yielding, besides the products, con-
densed fuel in the form of a mist and charred material in the
form of a smoke. Of course, the burning of hydrogen and oxygen
is simpler with no bluish tinge, smoke or char.

On the other hand, the burning of wood, which has been
studied extensively and can be used as a model for plastics,
is much more complex. Firstly, there must be a source of heat
energy, a thermal flux. This leads to degradation of the surface layers, yielding gases which are heated above their ignition temperatures and so ignite or take fire. This temperature is low for wood and higher for most plastics. The burning
gases yield heat which penetrates back into the wood and
degradation products continue to move out - heat transfer in
and mass transfer out. This process of combustion propagation
takes place at about 400°C. The temperature is higher, say
650°C at the char forming front. If the char remains behind
as a solid it is the familiar charcoal; if it breaks up into
small particles, it is smoke. If the degradation products
condense without burning they form a mist of liquids or tars.
This process continues until the wood is consumed. It may be
illustrated dramatically by the process of ablation [130,131].
Assuming that the gases evolved are fuels for combustion
ablation would be violent or explosive combustion.

The combustion of plastics under less dramatic conditions
is similar [132]. A pure hydrocarbon type such as polyethylene
produces carbon dioxide and water. Polyvinyl chloride tends
to form a char and liberate hydrogen chloride as well. Polyurethanes, polyisocryanurates, urea-formaldehyde polymers and
other thermosets, if they burn at all, usually yield a char.
There are many other variations from the normal process.
Fluoropolymers can be induced to burn in oxygen-rich atmos-
pheres and, in so doing, liberate fluorine atoms which react
with the surface layers giving energy which aids in the degradation [133]. Others tend to drip or to shrink away from a
flame [134]. In still others, as in wood, the char is sufficiently strong and incombustible to protect the polymer for

quite some time. The presence of combustible plasticizers or
fillers will promote combustion and the presence of inert and
incombustible fillers will tend to slow it down. On the other
hand, a filler may prevent a polymer from dripping or pulling
away from a fire and so maintain combustion of a film or fabric
which otherwise would not burn. A mixed fabric may be more
dangerous than a pure one. For example, cotton and polyester
are more combustible than either alone. The cotton ignites
easily; the polyester is then ignited but is unable to shrink
away because it is held by the cotton. The polyester burns
with a higher heat of combustion so it ignites more cotton and
the whole process is speeded up. However, heats of combustion
by themselves are not the governing factor.

The tests for combustion are numerous [135] and those used in
practice are for very mild conditions. A test for self-extinguishing characteristics, SE rating, means that the polymer,
when exposed to a flame for a short time, will not propagate
a flame. If it is retarded, the rate may be slower than that
for the pure polymer. However, in both cases, it must be
emphasized that the data are purely relative - if the temperature is raised high enough for long enough the material will
burn and may indeed reach a temperature where thermal decomposition will result in combustible gases being given off or
reach a temperature where an exothermic degradation of almost
explosive intensity may be observed. Improved tests to simulate the more severe service conditions are being evolved, but
in the meantime, while plastics are safer than wood, they cannot be treated with indifference.

Flame Retardancy

Just as there are numerous ways of combustion observed with
plastics, there are numerous ways of retarding the combustion.
Ignition may be slowed by using a heat reflecting surface,
often aluminum foil, or by using a surface with a high temperature of decomposition such as a halogenated polymer or a
fluoropolymer. In some cases the presence of a heat sink,
such as water, will assist, but, of course, it must be replaced
once it has been driven off. Then, during combustion, the

process may be altered to favor a char or a crust, which will
slow or stop the flow of heat in and of gases out. This process is normal for thermosets and may be increased by the char-
promoting additives, including fillers such as glass. Then,
the propagation may be stopped by the polymer dripping or
melting away, although this may not be a suitable method if
the drippings continue to burn and to spread the flame.
Finally, considerable success has been achieved in inhibiting
the glow of the char as for example, by using phosphates in
wood.

The usual method of inhibiting combustion or flame retar-
dancy is to add chlorinated or brominated materials, [137] often
with antimony oxide as a synergist, and inert fillers to
assist further. As mentioned earlier, the fillers tend to
form an inert surface and of course also promote char formation
and slow heat transfer and mass transfer. The halogen compounds act by forming stronger bonds on the surface, which
slows ignition even though the ignition temperature may be
reduced in some cases. The propagation step is slowed by a
complex mechanism whereby halogenated polymer is formed and
burns more slowly, both by an inhibition of the branching of
the oxidation reaction and by consuming energy to rupture the
stronger C-Cl bonds. The antimony oxide is believed to work
through conversion to antimony halide which halogenates the
polymer thereby perpetuating the effect of the halogen somewhat. The phosphate types tend to slow the glowing of char by
a process which is not well known. Whereas the CO formed on
the surface (26.43 K cal/mole) is normally converted to carbon
dioxide at or near the surface with enough heat (67.95 K cal/
mole) to propagate the charring reaction, this second step is
slowed in the presence of phosphate and the char goes out. In
general, the addition of flame retardants increases char and
smoke and decreases combustion, perhaps to the point where the
material may be regarded as retarded, self-extinguishing or
even non-combustible. However, it must be emphasized that in
more severe tests these results are misleading, the materials
will burn.

Flame retardants are of two main classes. The unreactive
ones are fillers added to the polymer during compounding, as

fillers or plasticizers. Often they weaken the polymer. This can be overcome by adding a reinforcing filler and, more particularly, by using a reactive flame retardant, which is part of the polymer, as for example, tetrabromophthalic anhydride or 2,2-bis(bromomethyl)-1,3-propandiol in the manufacture of polyesters. Other approaches, which do not destroy the properties of the polymers, are chlorination of polyethylene or of poly(vinyl chloride). Bromine is about twice as effective on a weight basis as chlorine, hence, preference for its use.

Among the many special treatments used in practice should be included intumescent paints which foam and char when heated and the foamed char insulates the substrate. This process is used extensively industrially but not so much in houses. Using cross-linked polymers or thermosets also helps but these may not foam as readily. Thus, there is often a peroxide or other curing agent present to simultaneously cross-link an intumescent mixture composed of a thermoplastic.

The final aim would be completely incombustible polymers [142] and there are many of these but they are as yet expensive and only used in such items as aircraft or firefighter's clothing, where they can stand temperatures up to about $1000^{\circ}C$. This is above most general fire conditions, but of course below that of a welding torch, etc. The basis of these polymers is a ladder or step-ladder structure which has many conjugated ring

Polybenzimidazole
step-ladder polymer

structures and very few or no hydrogen atoms. One of the types is fluorographite composed almost entirely of graphite and fluorine atoms. The conjugated ring structures absorb an enormous amount of energy and the lack of the evolution of any combustible gases affords no opportunity for combustion. The graphite will eventually glow as a char but the temperature for this to take place is very high.

While much has to be done to make plastics suitable for every use where fire might occur, it must be remembered that plastics are safer than natural products such as wood; they can be protected for casual or slight accidents, but only the most exotic can be used under extreme conditions. Someday these types will be available for use and already some of them are appearing as oven linings, pan linings, etc. Even some of the less exotic ones are appearing as plastic replacements for glass coffee makers and baking dishes.

The combustion of plastics also has its effect on incineration. At the moment the high fuel value of plastics is valued as an aid to incineration of garbage. As the combustibility is reduced, this effect will become less valuable particularly if it means also that more halogen, phosphorus and antimony may be in the flue gases. The use of plastics as land fill, along with other garbage, is not hazardous as far as we know since most plastics would remain undecomposed for centuries when buried to the normal depths, with the possible exception of poly(vinyl chloride) and some other types which require extensive stabilization. Recycling is another subject which is in an early stage of development except for the use of clean scrap in processing plants. The problems of recovery from complex mixtures arising from garbage offer real challenges. Pyrolysis of scrap tires and hydrolysis of scrap polyurethanes are processes which are in limited use to recover valuable residues.

References

1 G. Odian, Principles of Polymerization, McGraw-Hill, New York, 1970.

2 R.B. Seymour, Introduction to Polymer Chemistry, McGraw-Hill, New York, 1971.

3 K.J. Saunders, Organic Polymer Chemistry, Chapman and Hall, London, 1973.

4 D.B.V. Parker, Polymer Chemistry, Applied Science, London, 1974.

5 F. Rodriguez, Principles of Polymer Systems, McGraw-Hill, New York, 1970.

6 F.A. Bovey, Polymer Conformation and Configuration, Academic, New York, 1969.

7 F.W. Billmeyer Jr., Textbook of Polymer Science, Wiley-Interscience, New York, 2nd Ed., 1971.

8 D.J. Williams, Polymer Science and Engineering, Prentice-Hall, Englewood Cliffs, 1971.

9 P.A. Small, J. Appl. Chem., (London), 3(1953)71.

10 H. Burrell and B. Immergut in J. Brandrup and E.H. Immergut, (Eds.), Polymer Handbook, Wiley, New York, 1966, p. IV-341.

11 A. Beerbower, L.A. Kaye and D.A. Pattison, Chem. Eng., 74, No. 26(1967)118.

12 G. Gee, Trans. Inst. Rubber Ind., 18(1943)266.

13 A. Dobry and R. Boyer-Kawenoki, J. Polymer Sci., 2(1947)90.

14 R.A. Hayes, J. Appl. Polymer Sci., 5(1961)318, Rubber Chem. Technol., 35(1962)558

15 W.E. Wolstenholme, Polymer Eng. Sci., 8(1968)142.

16 J.R. Collier, Ind. Eng. Chem., 61 No. 9(1969)50.

17 R.F. Boyer, Polymer Eng. Sci., 8(1968)161.

18 R.F. Boyer, Rubber Chem. Technol., 36(1963)1303.

19 R.F. Boyer, J. Polymer Sci., C 14(1966)267.

20 R.F.Boyer, Plastics Polymers, 41(1973)15 and 71. (Reprint includes unpublished portion).

154

21 R.N. Haward (Ed.), The Physics of Glassy Polymers, Applied
   Sci., London, 1973.
22 J.M. McKelvey, Polymer Processing, Wiley, New York, 1962.
23 J.M. Schultz, Polymer Materials Science, Prentice-Hall,
   Englewood Cliffs, 1974.
24 P.H. Geil, Polymer Single Crystals, Interscience, New York,
   1963.
25 L. Mandelkern, Crystallization of Polymers, McGraw-Hill,
   New York, 1964.
26 B. Wunderlich, Macromolecular Physics. Crystal Structure,
   Morphology, Defects. Vol. 1, Academic, New York, 1973.
27 L.A. Wood and W. Bekkedahl, J. Res. Natl. Bur. Std.,
   36(1946)489; J. Appl. Phys., 17(1946)362; Rubber Chem.
   Technol., 19(1946)1145.
28 A. Peterlin (Ed.), Plastic Deformation of Polymers, Dekker,
   New York, 1971.
29 S.S. Voyutskii, Autohesion and Adhesion of High Polymers,
   Interscience, New York, 1963.
30 D.H. Kaelble, Physical Chemistry of Adhesion, Wiley-
   Interscience, New York, 1971.
31 D.F. Moore, The Friction and Lubrication of Elastomers,
   Pergamon, New York, 1972.
32 R.C. Bowes, W.C. Clinton and W.A. Zisman, Mod. Plastics,
   31(1954)131.
33 E.B. Atkinson, in P.D. Ritchie (Ed.), Physics of Plastics,
   van Nostrand, Princeton, 1965, p. 248.
34 R.S. Lenk, Plastics Rheology, Mechanical Behavior of Solid
   and Liquid Polymers, Wiley, New York, 1968.
35 S. Middleman, The Flow of High Polymers, Interscience, New
   York, 1968.
36 J.A. Brydson, Flow Properties of Polymer Melts, Iliffe,
   London, 1970.
37 R.E. Wetton and R.W. Whorlow (Eds.), Polymer Systems.
   Deformation and Flow, MacMillan, London, 1968.
38 A.S. Lodge, Elastic Liquids, Academic, New York, 1964.
39 C.S.W. Wells (Ed.), Viscous Drag Reduction, Plenum, New
   York, 1969.
40 M. Mooney and W.E. Wolstenholme, J. Appl. Phys.,
   25(1954)1098.

41 M. Mooney, J. Appl. Phys., 27(1956)691.

42 A.R. Berens and V.L. Folt, Trans. Soc. Rheol., 11(1967)95.

43 A.R. Berens and V.L. Folt, Polymer Eng. Sci., 8(1968)5.

44 A.N. Dunlop and H.L. Williams, J. Appl. Polymer Sci.,
   17(1973)2945.

45 R.P. White Jr., Polymer Eng. Sci., 14(1974)50.

46 J.D. Ferry, Viscoelastic Properties of Polymers, Wiley-
   Interscience, New York, (2nd Ed.) 1970.

47 J.J. Aklonis, W.J. MacKnight and M. Shen, Introduction to
   Polymer Viscoelasticity, Wiley-Interscience, New York, 1972.

48 R.M. Christensen, Theory of Viscoelasticity, An Introduction,
   Academic, New York, 1971.

49 I.M. Ward, Mechanical Properties of Solid Polymers, Wiley,
   New York, 1971.

50 L.E. Nielsen, Mechanical Properties of Polymers and
   Composites, Vol. 1, Dekker, New York, 1974.

51 P.I. Vincent, in P.D. Ritchie (Ed.), Physics of Plastics,
   van Nostrand, Princeton, 1965, p. 24.

52 J.R. McLoughlin and A.V. Tobolsky, J. Colloid Sci.,
   7(1952)555.

53 D. Kaplan and N.W. Tschoegl, Polymer Eng. Sci., 14(1974)43.

54 J.A. Sauer, SPE Trans., 2(1962)57.

55 H.K. Yip and H.L. Williams, Unpublished data, M.A.Sc.
   Thesis, 1974.

56 K.M. Sinnot, SPE Trans., 2(1962)65.

57 L.E. Nielsen, R.A. Wall and P.G. Richmond, SPE J.,
   11(1955)22.

58 B. Rosen (Ed.), Fracture Processes in Polymeric Solids,
   Phenomena and Theory, Wiley-Interscience, New York, 1964.

59 G.M. Bartenev and Yu.S. Zuyev, Strength and Failure of
   Viscoelastic Materials, Pergamon, New York, 1968.

60 G. Hulse, in P.D. Ritchie (Ed.), Physics of Plastics,
   van Nostrand, Princeton, 1965, p. 120.

61 M.N. Riddell, G.P. Koo and J.L. O'Toole, Polymer Eng. Sci.,
   6(1966)363.

62 S. Kase, J. Polymer Sci., 11(1953)425.

63 W. May, Trans. Inst. Rubber Ind., 40(1964)T109; Rubber
   Chem. Technol., 37(1964)826.

64 T.L. Smith, J. Polymer Sci., 32(1958)99.

65 T.L. Smith, and P.J. Stedry, J. Appl. Phys., 31(1960)1892;
   J. Polymer Sci., A-1(1963)3597.

66 T.L. Smith, J. Appl. Phys., 35(1964)27.

67 J.P. Berry and A.M. Bueche, in P. Weiss (Ed.), Proc. Symp.
   Adhesion Cohesion, Elsevier, Amsterdam, 1962, p. 18.

68 J.P. Berry, J. Polymer Sci., 50(1961)107 and 313.

69 D. Stefan and H.L. Williams, J. Appl. Polymer Sci.,
   18(1974)1279.

70 D. Stefan and H.L. Williams, J. Appl. Polymer Sci.,
   18(1974)1451.

71 D. Stefan, H.L. Williams, D.R. Renton and M.M. Pintar,
   J. Macromol. Sci., Phys., B4(1970)853.

72 M.D. Hartley and H.L. Williams, J. Appl. Polymer Sci.,
   in the press.

73 G. Kraus, Reinforcement of Elastomers, Wiley-Interscience,
   New York, 1965.

74 J.G. Mohr, S.S. Olessky, G.D. Shook and L.S. Meyer, SPI
   Handbook of Technology and Engineering of Reinforced
   Plastics/Composites, van Nostrand-Reinhold, Princeton, 1973.

75 L.E. Nielsen, Mechanical Properties of Polymers and
   Composites, Vol. 2, Dekker, New York, 1974.

76 L. Holliday, Composite Materials, Elsevier, Amsterdam, 1966.

77 D.M. Schwaber and F. Rodriguez, Rubber Plastics Age,
   48(1967)1081.

78 C. Markin and H.L. Williams, J. Appl. Polymer Sci.,
   18(1974)21.

79 L. Mullins, Rubber Chem. Technol., 20(1947)444, 21(1948)281.

80 E. Fischer and J.F. Henderson, Rubber Chem. Technol.,
   49(1967)1373.

81 N.K. Kalfoglou and H.L. Williams, J. Appl. Polymer Sci.,
   17(1973)1377.

82 R.S. Hunter and L. Boor, in J.V. Schmitz (Ed.), Testing of
   Polymers, Vol. 2, Interscience, New York, 1966, p. 279.

83 J.J. Gouza, in J.V. Schmitz (Ed.), Testing of Polymers,
   Vol. 2, Interscience, New York, 1966, p. 225.

84 P.E. Slade Jr. and L.T. Jenkins, Thermal Analysis, Dekker,
   New York, 1970; Thermal Characterization Techniques, Dekker,
   New York, 1970.

85 M. Gordon, in P.D. Ritchie (Ed.), Physics of Plastics,

van Nostrand, Princeton, 1965, p. 209.

86 B. Ke (Ed.), Thermal Analysis of High Polymers, J. Polymer Sci. C. Polymer Symposia. No. 6, Wiley-Interscience, New York, 1964.

87 R.H. Norman, Conductive Rubbers and Plastics, Their Production, Application and Test Methods, Elsevier, Amsterdam, 1970.

88 A.T. MacPherson, Rubber Chem. Technol., 36(1963)1230.

89 J.E. Katon, Organic Semiconducting Polymers, Dekker, New York, 1968.

90 L. Boguslavskii and A.V. Vannikov, Organic Semiconductors and Biopolymers, Plenum, New York, 1970.

91 N. Parkman, in P.D. Ritchie (Ed.), Physics of Plastics, van Nostrand, Princeton, 1965, p. 285.

92 P.F. Bruins (Ed.), Plastics for Electrical Insulation, Wiley-Interscience, New York, 1968.

93 A.H. Sharbaugh, in J.V. Schmitz (Ed.), Testing of Polymers, Vol. 1, Interscience, New York, 1965, p. 201.

94 A.H. Scott, in J.V. Schmitz (Ed.), Testing of Polymers, Vol. 1, Interscience, New York, 1965, p. 213.

95 R.W. Tucker, in J.V. Schmitz (Ed.), Testing of Polymers, Vol. 1, Interscience, New York, 1965, p. 237.

96 R.W. Warfield, in J.V. Schmitz (Ed.), Testing of Polymers, Vol. 1, Interscience, New York, 1965, p. 271.

97 T.W. Dakin, in J.V. Schmitz (Ed.), Testing of Polymers, Vol. 1, Interscience, New York, 1965, p. 297.

98 W.D. Kingery, Introduction to Ceramics, Wiley, New York, 1960, p. 647, 686.

99 W.A.Little, in A. Rembaum and R.F. Landel (Eds.), Electrical Conduction of Polymers, J. Polymer Sci. C, Polymer Symposia No. 17, Wiley-Interscience, New York, 1967, p. 3.

100 C.F. Chen and H.L. Williams, Brit. Polymer J., 5(1973)183.

101 M. Vernois and H.L. Williams, Can. J. Chem., in the press.

102 J. Haslam, H.A. Willis and D.C.M. Squirrell, Identification and Analysis of Plastics, Iliffe, London, (2nd Ed.), 1972.

103 D.W. Saunders, in P.D. Ritchie (Ed.), Physics of Plastics,

van Nostrand, Princeton, 1965, p. 387.

104 J.T. Yang, in B. Ke (Ed.), Newer Methods of Polymer
Characterization, Wiley-Interscience, 1964, p. 103.

105 P. Crabbe, Optical Rotatory Dispersion and Circular Dichroism in Organic Chemistry, Holden-Day, San Francisco, 1965.

106 R.S. Stein, Rheo-optics of Polymers, J. Polymer Sci. C.
Polymer Symposia. No. 5, Wiley-Interscience, New York, 1964.

107 R.S. Stein, in B. Ke (Ed.), Newer Methods of
Polymer Characterization, Wiley-Interscience, 1964,
p. 155.

108 E.F. Gurnee, L.T. Patterson and R.D. Andrews, J. Appl.
Phys., 26(1955)1106.

109 J. Durisin and H.L. Williams, J. Appl. Polymer Sci.,
17(1973)709.

110 J.F. Henderson, K.H. Grundy and E. Fischer, J. Polymer
Sci., C. 16(1968)3121.

111 H. Janeschitz-Kriegel, Advances in Polymer Sci.,
6(1969)170.

112 V.N. Tsvetkov, in B. Ke (Ed.), Newer Methods of Polymer
Characterization, Wiley-Interscience, New York,
1964, p. 563.

113 B.M. Murphy, Chem. Ind., London, 10(1969)289.

114 R.W.B. Stephens, in P.D. Ritchie (Ed.), Physics of
Plastics, van Nostrand, Princeton, 1965, p. 410.

115 C.F. Chen and H.L. Williams, unpublished data, Ph.D.
Thesis, 1974.

116 M.P. Blake and W.S. Mitchell (Eds.), Vibration and
Acoustic Measurement Handbook, Spartan, New York, 1972.

117 C.F. Chen and H.L. Williams, unpublished data, Ph.D.
Thesis, 1974.

118 J.A. Grates, J.E. Lorenz, D.A. Thomas and L.H. Sperling,
Mod. Paint Coatings, 65, No. 2(1975)35.

119 Yu.S. Urzhumtsev and S.L. Skalozub, Polymer Mech.,
5(1969)88.

120 P.C. Bruel and H.P. Olsen, Bruel & Kjaer Tech. Rev.,
No. 3(1973)14.

121 N. Grassie, Chemistry of High Polymer Degradation
Processes, Interscience, New York, 1956.

122 H.H.G. Jellinek, Degradation of Vinyl Polymers, Academic,

New York, 1955.

123 L. Reich and S.S. Stivala, Elements of Polymer Degradation, McGraw-Hill, New York, 1971.

124 S.L. Madorsky, Thermal Degradation of Organic Polymers, Wiley-Interscience, New York, 1964.

125 R.T. Conley, Thermal Stability of Polymers, Vol. 1, Dekker, New York, 1970; Vol. 2, Dekker, New York, 1974.

126 O. Vogl, H.C. Miller and W.H. Sharkey, Chem. Eng. News, Mar. 20 (1972)41.

127 W.L. Hawkins, Polymer Stabilization, Wiley-Interscience, New York, 1972.

128 L. Reich and S.S. Stivala, Autoxidation of Hydrocarbons and Polyolefines, Kinetics and Mechanism, Dekker, New York, 1969.

129 E. Weiss, in J.V. Schmitz (Ed.), Testing of Polymers, Vol. 2, Interscience, New York, 1966, p. 349.

130 G.F. D'Alelio and J.A.Parker, Ablative Plastics, Dekker, New York, 1971.

131 L. Schmidt, Mod. Plastics, 38, No. 3(1960)131; No. 4 (1960)147.

132 P.C. Warren, SPE J., 27, No. 2(1971)17.

133 C.P. Fenimore and G.W. Jones, J. Appl. Polymer Sci., 13(1969)285.

134 E.V. Gouinlock, J.F. Porter and R.R. Hindersinn, Pre-prints, Div. Org. Coatings Plast. ACS, 28(1968)225.

135 L.B. Allen and L.N. Chellis, in J.V. Schmitz (Ed.), Testing of Polymers, Vol. 2, Interscience, New York, 1966, p. 349.

136 Flame Resistance with Plastics, Conference Supplement No. 2, The Plastics Institute, Transactions and Journal, Pergamon, London, Jan. 1967.

137 A.D. Delman, J. Macromol. Sci. Rev. Macromol. Chem., C3(1969)281.

138 P. Thiery, Fireproofing, Chemistry, Technology and Applications, Elsevier, Amsterdam, 1970.

139 J.W. Lyons, The Chemistry and Uses of Fire Retardents, Wiley-Interscience, 1970.

140 K.C. Salooja, Advances in Chem., 76(1968)226.

141 J.A. Schneider, R.G. Pews and J.D. Herring, Preprints.

Div. Org. Coatings Plast. ACS, 29(1969)382.
142 N. Platzer, Ind. Eng. Chem., 61, No. 5(1969)10.

Subject Index

162

166